Steam Turbines
and
Steam Power Plant

Steam Turbines and Steam Power Plant

R. Jaswal
Head,
Department of Mechanical Engineering
Suresh Gyan Vihar University
Jaipur

Dr. R.K. Purohit (Retd.)
Associate Professor
Dept. of Mechanical Engineering,
M.B.M. Engineering College,
JNV University, Jodhpur

SCIENTIFIC PUBLISHERS

Published by

SCIENTIFIC PUBLISHERS (INDIA)
5 A, New Pali Road, P.O. Box 91
Jodhpur 342 001 (INDIA)
E-mail: info@scientificpub.com
Website : www.scientificpub.com

Print: 2020

ISBN: 978-81-72337-61-2
eISBN: 978-93-87741-45-4

Printed in India

SYLLABUS

Rajasthan Technical University (RTU)

STEAM TURBINES AND STEAM POWER PLANT

Unit 1. Steam Turbines

Principle and working of steam turbines, type of turbines, impulse and reactions, compounding for pressure and velocity. Velocity triangles for various types.

Unit 2.

Stage efficiency, diagram efficiency, steam speed to blade, speed ratio for optimum performance. Energy losses in steam turbine, turbine performance at various leads and governing of steam turbines. Construction details and description of steam turbine components in brief.

Unit 3: Regenerative feed heating cycles

Introduction: Most Ideal Regenerative feed heating cycle. Regenerative feed heating cycles and their representation of T-s and h-s Diagram. Representation of actual process of T-s and h-s Diagram Regenerative cycles. Other types of feed heating arrangements. Optimum feed water temperature and saving in Heat Rate. Feed Heaters, Direct Contact Heaters, Surface Heaters.

Reheating – Regenerative and Regenerative water – Extraction Cycles.

Reheating of steam, Practical reheating and Non-reheating cycles, advantage and disadvantages of reheating, regenerative water extraction cycles, practical heating arrangements.

Unit 4

Governing and performance of Steam Turbines.

Description of back pressure Turbines, pass-out Turbines and Mixed Pressure Turbines.

Unit 5: Steam Power Plant

Steam power plant selection of location, working medium, Fuels and fuel handling equipments, ash handling equipments. Air pre-heater, feed water treatment. Methods of combustion and various type of combustors. Types of boilers, Modern developments in steam boilers. Description of cooling tower.

PREFACE

It gives me immense pleasure in placing this book `Steam Turbine and Steam Power Plant' in the hands of B.Tech VII Semester students of Mechanical Engineering of Rajasthan Technical University (RTU).

The syllabus of steam turbine and steam power plant is divided into Five units; and likewise this book is also divided into Five Chapters. The subject matter has been evolved from first principles and the treatment is then extended to considerable depth so that this single book covers the complete course of VII Semester of RTU.

The importance of illustrative examples cannot be under estimated, as they help in understanding the concepts of the subject clearly. Besides, they also follow the style of question set in RTU.

We are thankful to colleagues, friends and students who encouraged us to write this book.

We express our appreciation and gratefulness to our publisher Tanay Sharma for his most cooperative, painstaking attitude and untiring effort in bringing out this book in a very short period.

We have used several references and we are grateful to those learned authors.

R. Jaswal
Dr. R.K. Purohit

CONTENTS

Chapter -1

STEAM TURBINES: PRINCIPLE OF WORKING AND VELOCITY TRIANGLES

1.1. INTRODUCTION

The steam turbine was invented in the last decade of the nineteenth century. It has undergone several changes in design during the past nine decades. The steam turbine has been used predominantly as prime-mover in all thermal power stations. It is not likely to be replaced in the foreseeable future. Now-a-days, single steam turbines of 1000 MW capacity are built in many countries and unit of 1500 MW capacity is planned in future power programmes continuing developments in materials and other areas promise even better performance and low cost.

In steam turbine the working fluid enters at a high pressure and acquires increased kinetic energy as it expands to a lower pressure in a ring of fixed nozzles. The stream of fluid then undergoes a change of momentum as it flows through passage between blades attached to the rotor, and component in a direction tagential to the circle of rotation produces the output torque at the shaft. In the turbines the fluid undergoes a continuous steady flow process and the speed of flow is very high. For this reason comparatively small machines can handle large mass flows and large work transfers, and the processes can be assumed to be adiabatic. Where as in the reciprocating steam engine, the pressure of steam is used to overcome external resistance and dynamic action of the steam is negligibly small.

Steam turbines offer many advantages over steam engine as prime movers. Some of these are listed below.

(1) The specific steam consumption does not increase with years of service.

(2) The steam turbine can take considerable overload with only a slight reduction in its efficiency.

(3) The vibrations and noise are greatly minimised due to the absence of reciprocating parts.

(4) Flywheel is not needed. Balance the turbine rotor very accurately. Reduces the cost of foundation.

(5) There is no condensation loss as parts of the steam turbines are subjected to constant temperature and pressure.

(6) Condensate can be used directly in the boiler without pretreatment as there is very little chance of lubricating oil mixing with steam.

(7) The maintenance cost is low because of the highly simplified construction. The life of the prime mover is increased considerable.

(8) Higher speeds are possible as only rotary motion is used. The specific steam consumption is considerably reduced.

(9) The thermodynamic efficiency of a turbine plant is higher than that of the steam engine.

(10) Steam engine works on modified Rankine cycle and steam turbine works on Rankine cycle. The expansion of steam can be continued upto the lowest pressure possible thus obtaining more work for the same pressure ratio as compared to steam engine.

1.2 CLASSIFICATION OF STEAM TURBINES

Steam turbines are classified as `impulse' or `reaction' types. In fact, pure reaction steam turbines are not manufactured, and existing so called `reaction' turbines are therefore better called `impulse-reaction' types because their operation is partly due to impulse and partly due to reaction.

Steam turbines can also be classified according to the direction of steam flow as

(1) Axial flow turbine, and

(2) Radial flow turbine

Mostly the steam turbines are of axial flow type (impulse as well as impulse reaction type) but only radial flow steam turbine (L jungstorm turbine) is a reaction type.

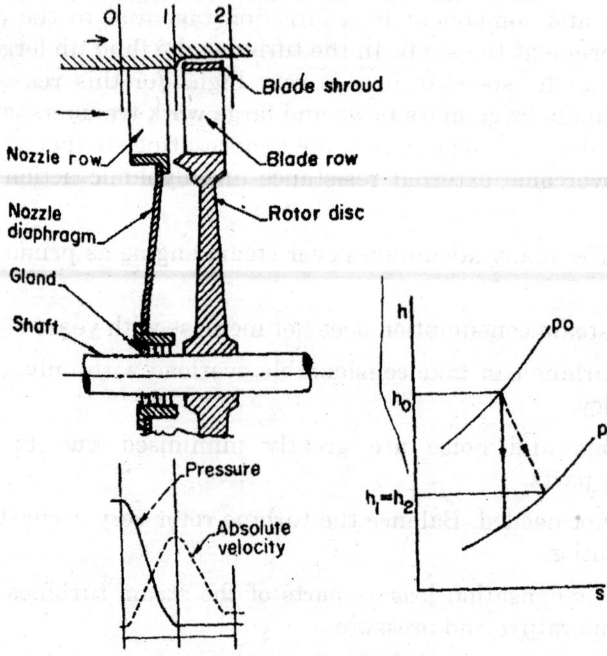

Fig. 1.1 Impulse Turbine

1.2.1 Impulse turbine

In an impulse turbine all the available pressure drop from supply to exhaust occurs across the nozzles so that the steam carries out its full expansion in the nozzles and emerges with high velocity. The nozzle is direction so that the steam glides on to the blades, and these are so shaped that the direction of flow steam is changed and thereby a force is exerted on the blades (Fig. 1.1). The examples of impulse turbine are De-Laval. Curties and Rateau.

1.2.2 Impulse - reaction turbine

In an impulse-reaction turbine, only part of the pressure drop occurs in the nozzles, the remainder occurring during the passage of the steam through the blades. The blade passage are nozzle shaped so that the acceleration of the steam occurs partly in the nozzles, and partly in the blades. Since acceleration requires a force, a resultant reaction occurs on the blades. The force of reaction is added to the force resulting from redirection of the steam to make the total propulsive force on the blades. In most modern reaction turbines the propulsive force is derived from 50% reaction and 50% impulse effect (Fig. 1.2) and Fig. 1.2b shows types of blades.

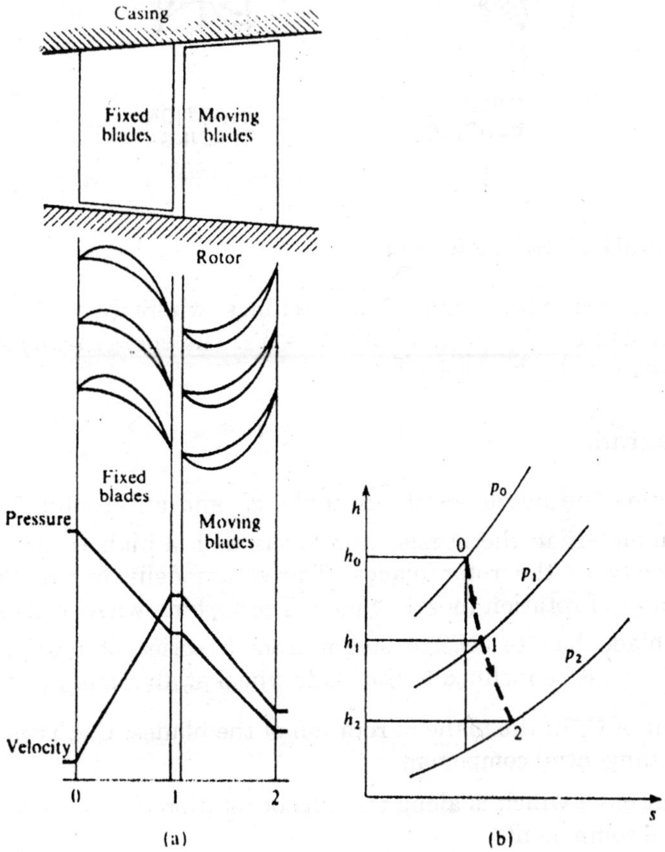

Fig. 1.2(a). Impulse reaction turbine

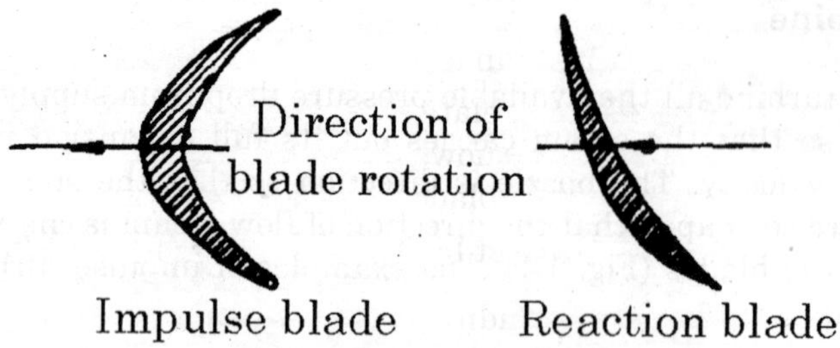

Fig. 1.2.b. Types of blade.

In an impulse turbine, nozzles are arranged in the turbine casing, whereas in the reaction turbine a ring of fixed blades acts as the nozzles for each stage (Fig. 1.3).

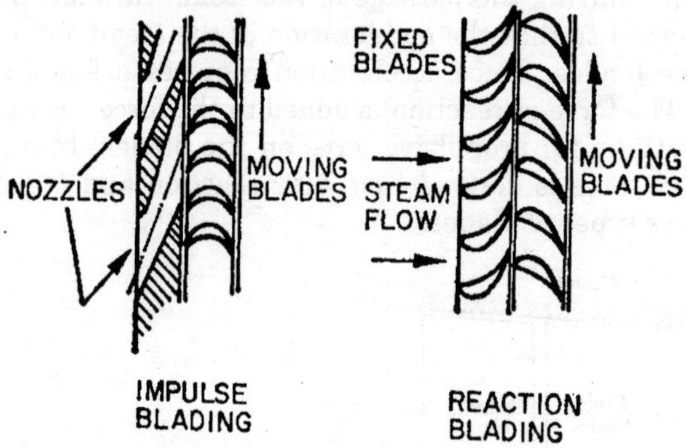

Fig. 1.3. Impulse and impulse-reaction turbine blading

1.3 THE SIMPLE IMPULSE TURBINE

Impulse turbines are mostly axial flow machines, where $U_1 = U_2$, and since these, machines have zero degree of reaction, hence $V_1 = V_2$. The energy transfer thus is wholly derived from a change of absolute velocity.

1.3.1 Velocity diagram

The steam enters the nozzles with an angle α_0 and an absolute velocity C_0. The steam expands completely in the nozzles and levels with a high velocity C_1. This is the absolute inlet velocity to the rotor blades. The steam delivered to the wheel at an angle α_1 with the axis of rotation (nozzle angle). The turbine wheel consists of a series of blades and each blade has to receive steam from a series of nozzles in succession, therefore the nozzle must be inclined to the blade wheel at an angle α_1.

The component of C_1 in the plane of rotation of the blades, C_{w1}, is called the velocity of whirl at inlet or tangential component.

The component of C_1 which is along the axis of rotation, C_{a1}, is called the velocity of flow at inlet or axial component.

Now, in operation, the turbine blades will be rotating. Let the mean blade speed be U. The mean blade speed will be the speed at the mean height of the blades.

Since the steam has velocity C_1 at an angle α_1 and the blade is moving with velocity U, then, the velocity of the steam relative to the blade, V_1, will be obtained by compounding these two velocities as shows in the inlet velocity triangle. V_1 will be inclined to the plane of rotation of the blades at angle β_1. If the steam is to enter the blades without shock, then angle β_1 must be the inlet angle of the blades.

Now consider the outlet or exit conditions. The steam at inlet will have an ongoing velocity relative to the blade of V_1. As the steam passes over the blade its direction will be changed and it will leave the blade with an exit relative velocity of V_2 at an angle β_2 to the plane of rotation, which is the exit angle of the blade. However if there is friction the relative velocity at outlet will be lower than the relative velocity at inlet (i.e. $V_1 > V_2$). The relative velocity at outlet considering friction is given by

$$V_2 = K\,V_1 \tag{1.1}$$

where, K is called the blade velocity coefficient.

The steam at outlet will therefore have two component velocities, namely, V_2 and U. The absolute velocity, C_2, will be obtained by compounding these two velocities.

Let α_2 be the angle made by the absolute velocity with the axis of rotation. Again, the absolute velocity at outlet of the moving blades has two components, the whirl component C_{w2} and the flow component C_{a2}.

From the above information the velocity triangles can be drawn at inlet and outlet of the moving blades (Fig. 1.4).

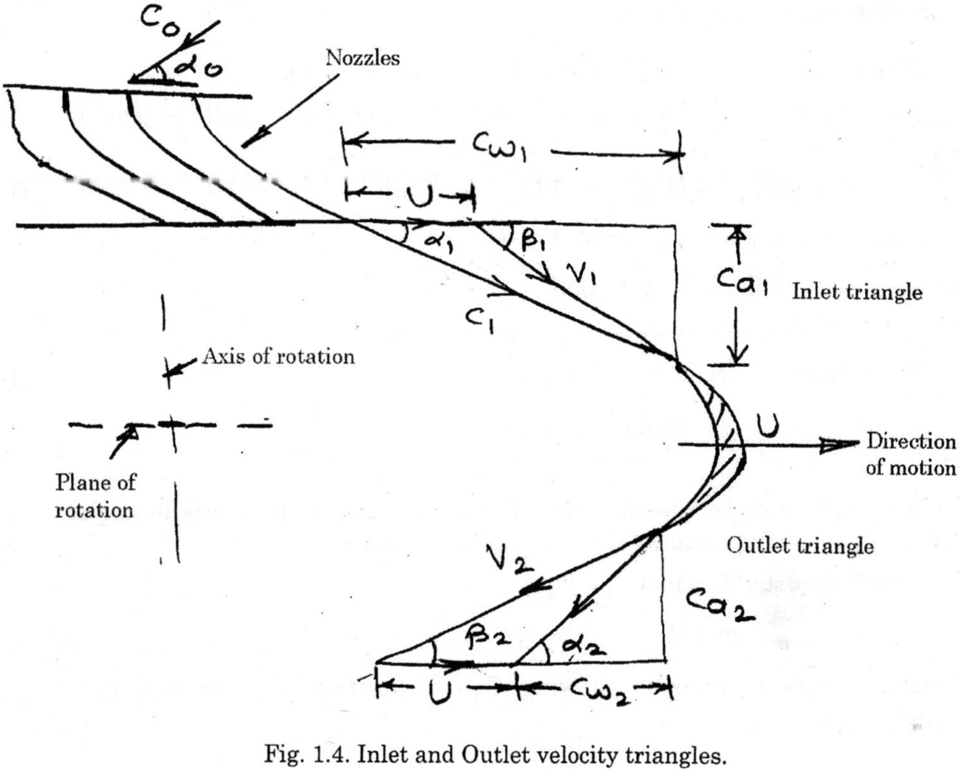

Fig. 1.4. Inlet and Outlet velocity triangles.

1.3.2 The combined inlet and outlet velocity triangles

Since both the inlet and outlet velocity triangles have a common vector, U, the mean blade speed, then they can be combined into a single diagram as shown in Fig. 1.5.

It is to be noted that both absolute velocities start from A. Both relative velocities start from B. If there is no friction in the blade then $V_2 = V_1$ and hence the extremities of both relative velocities lie on the arc of a circle, centre B.

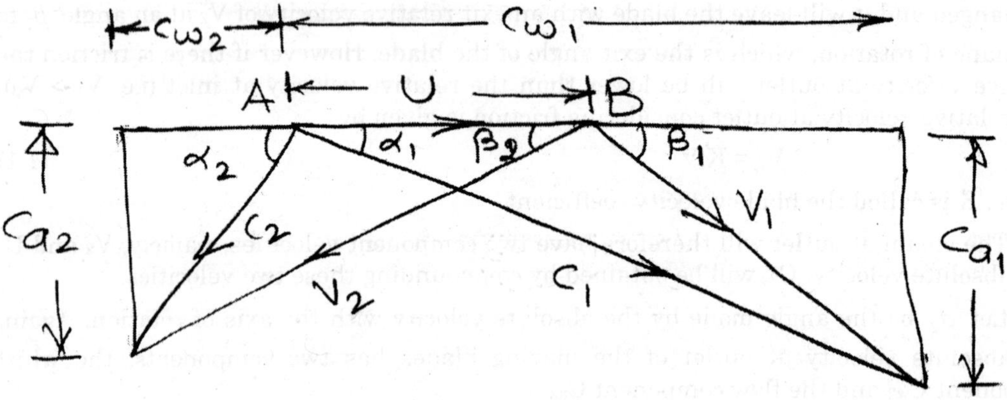

Fig. 1.5. Combined inlet and outlet velocity triangles.

1.3.3 Work done and axial thurst

According to Newton's second law of motion we have:

The rate of change of momentum is proportional to the externally applied force and takes place in the direction in which the force acts.

$$F = \dot{m}\,(C_{w1} - C_{w2}) = \dot{m}\ \Delta\ C_w \tag{1.2}$$

This is known as tangential thrust. F = F_t.

Now work done on blades per second.

= Force x Distance move per second.

$$\therefore \text{Work done per second} = \dot{m}\ \Delta\ C_w\ U \tag{1.3}$$

$$\therefore \text{Power developed} = \frac{\dot{m}\,\Delta\,C_w\ U}{1000}\ (\text{kW}) \qquad (1.4$$

It is now necessary to consider the change of momentum at right angles to the blades since any resulting force would cause an axial thrust.

$$\therefore \text{Axial thrust} = F_a = \dot{m}\,(C_{a1} - C_{a2})$$

$$= \dot{m}\,\Delta\,C_a \tag{1.5}$$

A combined velocity triangle showing C_{w2} is positive. C_{w2} is zero and C_{w2} is negative. (Fig. 1.6)

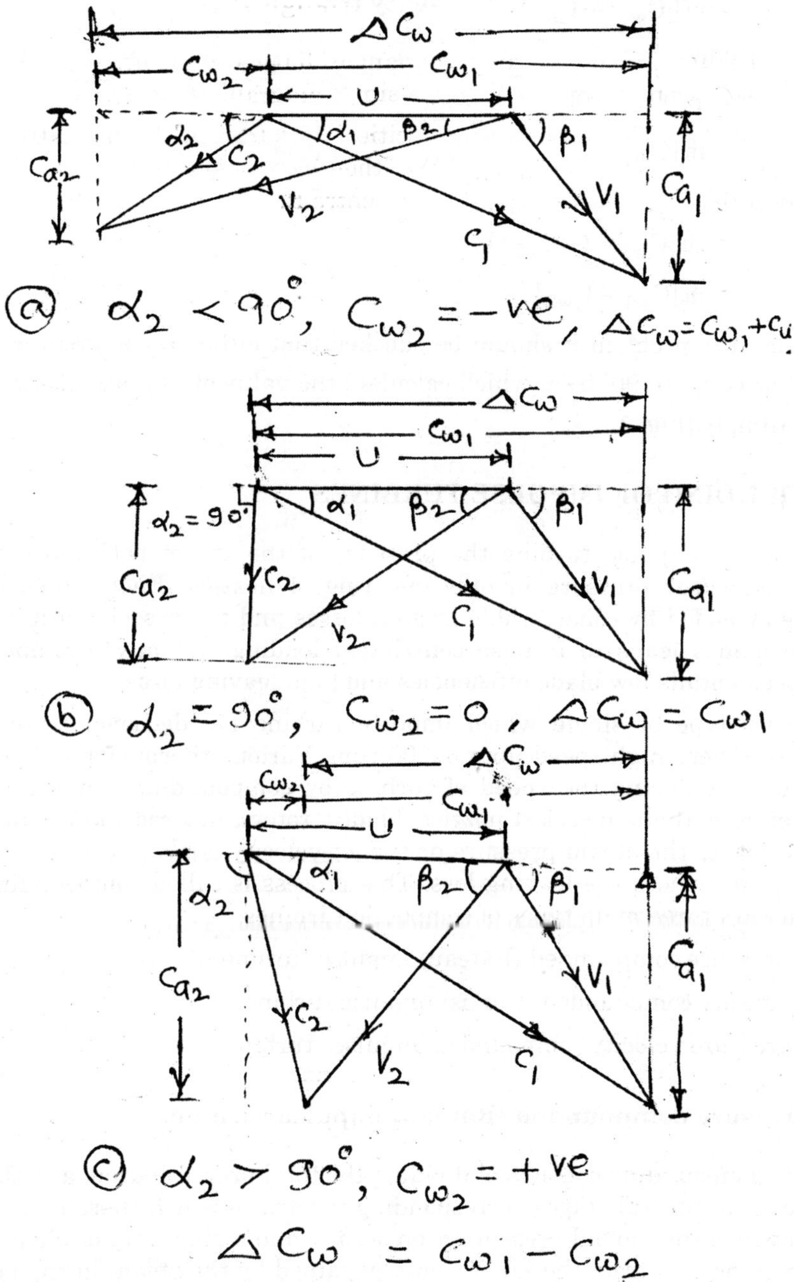

Fig. 1.6. Combined velocity triangle showing the effect of outlet whirl component.

Using equation (1.2) for triangles 1.6(a), 1.6(b) and 1.6(c), we get.

From Fig. 1.6 (a)

$$F = \dot{m}\,[\,C_{w1} - (-C_{w2})]$$

$$= \dot{m}[C_{w1} + C_{w2}] \quad (1.7)$$

From Fig. 1.6(b)

$$F = \dot{m}[C_{w1} - 0]$$
$$= \dot{m}[C_{w1}]$$

From Fig. 1.6(c)

$$F = \dot{m}[C_{w1} - (+C_{w2})]$$
$$= \dot{m}[C_{w1} - C_{w2}]$$

For solving the problem it should be checked that either α_2 is greater than 90° or less than 90° or equal to 90° from which calculate the value of C_{w2} and draw the velocity triangle accordingly [Fig. 1.6].

1.4. COMPOUNDING OF IMPULSE TURBINES

In the simple impulse turbine the pressure of the steam is dropped from boiler pressure to condenser pressure in only one ring of nozzles. Enormously high steam velocities are attended by considerable friction losses and the associated high rotational speed of the blades leads to intense centrifugal loading, yet any attempt to restrict rotational speed entails low blade efficiencies and high leaving losses.

The earlier type of single wheel impulse turbine i.e. de-Lavel turbine had the disadvantages of very high speed upto 30000 rpm. Various thermodynamic means have been adopted for reducing the speed of turbine by compounding. This is achieved by making use of more than one set of nozzles, blades, rotors, in series, keyed to a common shaft, so that either the steam pressure or the jet velocity is absorbed by the turbine in stages. This also reduces the leaving loss. This process is called compounding of steam turbine. There are three main types of compound turbines.

(a) The pressure compounded (Rateau) impulse turbine.

(b) The velocity compounded (Curtis) impulse turbine.

(c) The pressure velocity compounded impulses turbine.

1.4.1 The pressure compounded (Rateau) impulse turbine

Pressure compounding consists of dividing the total pressure drop, and therefore the enthalpy drop into several stages corresponding to each to which there is a separate set of nozzles in which the partial pressure drop occurs, and a single set of blades connected to a common rotor to extract the kinetic energy gained by the steam in the nozzles. It is usual for the pressure drop in each stage to be adjusted to give the same enthalpy drop per stage, so that the same power is developed in each stage (Fig. 1.7).

Since only part of the pressure drop occurs in each stage, then the steam velocities will not be so high and hence the turbine will run slower. All stages, however, one coupled to the same shaft, with the result that there is no loss of output.

This is the most efficient type of turbine because the ratio of blade velocity to steam velocity remains constant, but it has the disadvantages of large number of stages; hence it is most expensive. Therefore, pressure compounded turbine has now become out of date.

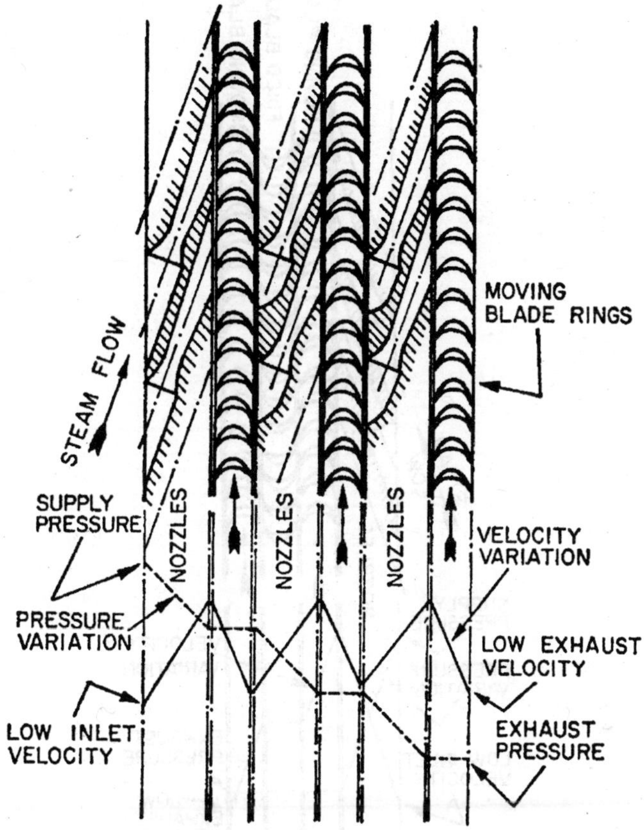

Fig. 1.7 Pressure compounding - three stages.

1.4.2 The velocity compounded (Curtis) impulse turbine

In this type the total heat drop takes place in the nozzles, but the velocity generated is utilised in two or more rings of blades. All the expansion takes place in a single set of nozzles, and the steam then passes through a series of blades attached to a single wheel or rotor. Since the blades move in the same direction it is necessary to change the direction of the steam between one set of moving blade and the next. For this purpose a stationary ring of blades is fitted between each pair of moving blades. A two row wheel version of the turbine is shown in Fig. 1.8.

Since only part of the velocity of steam is used in each ring of blades the speed of the turbine will be restricted. It is also to be noted that all pressure drop occurs in the nozzles, the pressure remaining constant over each ring of moving and fixed blades.

The advantages of velocity compounded turbine is relatively fever number of stages and hence less initial cost. However, the disadvantages is low efficiency as the ratio of blade velocity to steam velocity is not optimum for all the wheels. Therefore, efficiency goes on decreasing for subsequent stages. Hence, large velocity compounded turbines are obsolute.

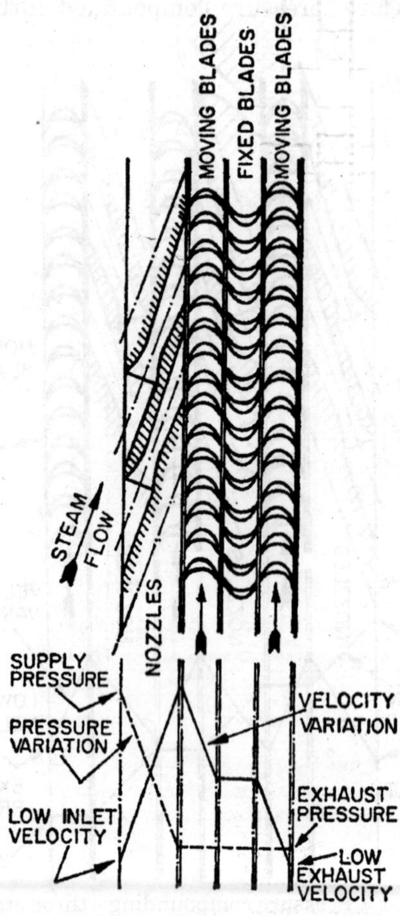

Fig. 1.8. Velocity compounding - two stages.

1.4.3 Pressure velocity compounded turbine

This turbine is a combination of pressure - velocity compounding and is shown diagrammatically in Fig. 1.9.

Pressure velocity compounding was an effort in the direction of combining the advantages of previous two types. The total pressure drop of the steam is divided into stages and the velocity obtained in each stage is also compounded. It results in bigger pressure drop in each stage and hence less number of stages which means a more compact turbine than a pressure compounded turbine. However, the efficiency is lower than a pressure compounded turbine.

Compared to the pressure compounded impulse turbine this arrangement was more popular due to its simple construction, but because of its low efficiency it is rarely used now-a-days.

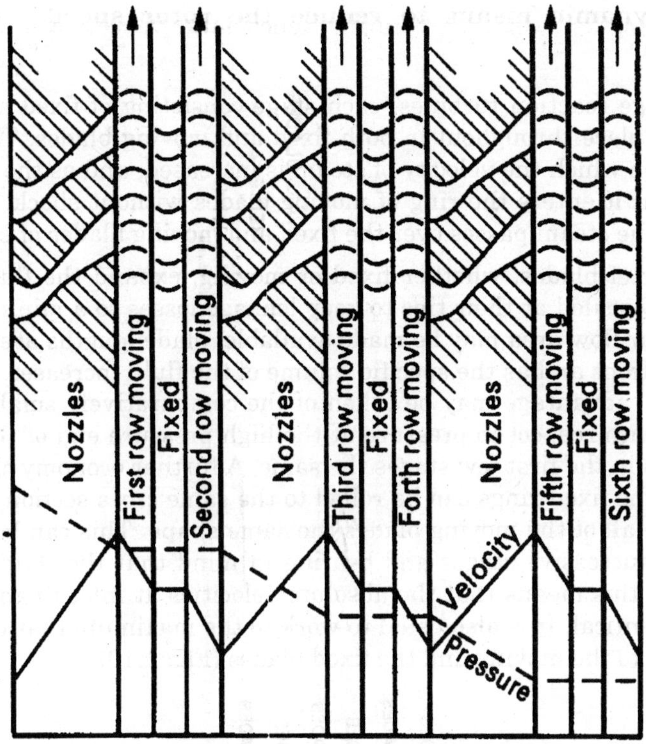

Fig. 1.9. Pressure - Velocity compounding

1.5 IMPULSE-REACTION TURBINE

The exact opposite of the impulse turbine would be a full reaction turbine. Pure reaction is impractical and so called reaction turbines are really mixed impulse - reaction turbines in which the drums is driven round partly by absorbing the kinetic energy of the steam emerging from the casing blades and partly by reaction from the pressure drop occurring in the moving blade passages.

In the impulse-reaction turbine the pressure drop occurs over the moving blades so that the steam expands continuously and its relative velocity at outlet is greater than its relative velocity at inlet. Fixed casing blades replaces the nozzles of the impulse turbine, and pressure drop occurs both in the fixed casing blades and the moving blades (Fig. 1.2).

A stage of a impulse - reaction turbine (reaction turbine) consists of a fixed row of blades over the whole of the circumferential annulus and equal number of moving blades on a wheel. Admission of steam in the reaction turbine take place over the complete annulus, i.e. there is full admission. The pressure drop takes place both on fixed and moving blades, therefore the blade passage in both are of convergent nozzle shape.

The reaction turbine was invented by Sir Charles Parsons who produced a 7.5 kW turbine running at 17000 rpm in 1884. Parsons was a contemporary of de Laval. These turbines are not manufactured now a days because of very high speeds.

1.5.1 Thermodynamic means to reduce the rotor speed - compounding of reaction turbine

In multi-stage reaction turbines, each stage consisting of fixed and moving blades. Heat drop takes place throughout in both fixed and moving blades. The fixed blades act both as nozzles, in which the velocity of steam is increased, and as the means of directing the steam so that it enters the ring of moving blades without shock. The pressure falls continuously as the steam passes over the fixed and moving blades of each stage.

As each row of blades, whether fixed or moving, expand the fluid, then all blades must be carefully sealed at their tips to keep leakage losses to a minimum. In order that a suitable annular flow area may be made available, blade lengths are steadily increased as the pressure drops and as the specific volume of the fluid increases along the length of the turbine. The advantage may be taken of the comparatively small rate of change of specific volume with respect to pressure at the high pressure end of the turbine to make the blade heights in the first few stages the same. A further economy can be effected if all of the blades in the fixed rings can be rolled to the same cross section, and likewise, it is desirable to have all of the moving blades the same shape. This can be done if uniformly of flox exists in successive stages and bearing in mind that the fluid leaving one stage enters the next - this means that the absolute velocities at inlet to and outlet from each stage must be identical. It is also usual to work to the maximum that the axial velocity is the same at inlet to the moving and the fixed blades (Fig. 1.10).

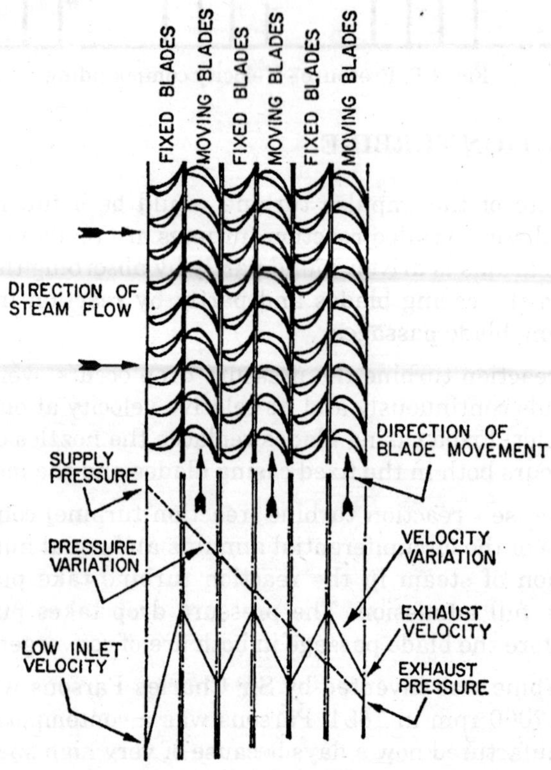

Fig. 1.10 Compounding of reaction-turbine

It is no longer convenient to talk of 'nozzles' and 'blades', since in the reaction turbine both fixed and moving blades act as nozzles. It is usual to refer to the two sets of blades as the stator blades and the rotor blades.

1.5.2 Degree of reaction and velocity diagram of reaction turbine

The degree of reaction is a measure of the proportion of the work done by reaction effect and may be defined as

$$R = \frac{\text{Enthalpy drop in moving blades}}{\text{Enthalpy drop in stage}}$$

$$= \frac{\text{Enthalpy drop in moving blades}}{\text{Enthalpy drop in moving blade} + \text{Enthalpy drop in fixed blade}}$$

$$= \frac{h_1 - h_2}{h_0 - h_2} \tag{1.9}$$

where, h_0, h_1 and h_2 are the enthalpies of the steam at inlet to the fixed blades, at entry to the moving blades, and at exit from the moving blades respectively.

In the Parsons impulse-reaction turbine, "half degree" or 50% reaction is used so that equal proportions of energy are contributed by the impulse and reaction velocity in the moving blades exactly equals the increase in absolute velocity in the fixed blades. Hence fixed and moving blades must be made identical, and as a result the velocity triangle at outlet is identical in shape to that at inlet but in reversed direction, therefore, $\alpha_1 = \beta_2$, $\alpha_2 = \beta_1$, $C_1 = V_2$ and $C_2 = V_1$ (Fig. 1.11).

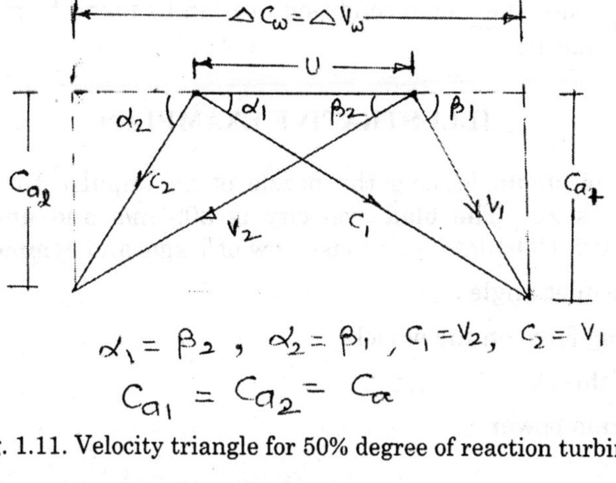

Fig. 1.11. Velocity triangle for 50% degree of reaction turbine.

1.5.3 Height of blades of a reaction turbine

In reaction turbines which "flow full", the volume of steam flowing increases due to continual reduction in pressure, and the blade heights have to increase progressively to accommodate it. Steam fills the entire annual area between the base and shroud of the turbine blading apart from the area occupied by the blade section itself (Fig. 1.12).

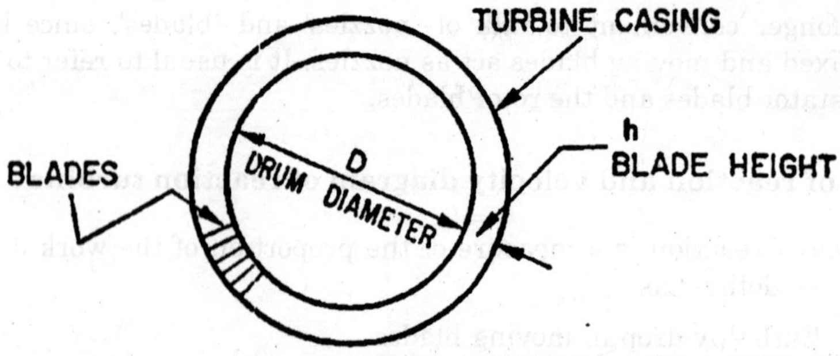

Fig. 1.12. Blade height

Annular area through steam flows

$$= \frac{\pi}{4}(D + 2h)^2 - \frac{\pi}{4}D^2$$

$$= \pi h (D + h)$$

Volume of steam flowing per second through the turbine

$$= \text{Annular area} \times \text{Axial velocity of flow}$$

$$= \pi h (D + h) \times C_a$$

The mass of steam flowing per second

$$\dot{m} \times v = \pi h (D + h) C_a$$

Therefore,
$$\dot{m} = \frac{\pi h (D + h) C_a}{v} \quad (1.10)$$

At any reaction pair, the steam condition (x) can be found from h-s chart, and hence also the specific volume v.

ILLUSTRATIVE EXAMPLES

1.1. The velocity of steam leaving the nozzle of an impulse turbine is 900 m/s and nozzle angle is 20°. The blade velocity is 300 m/s and and the blade velocity coefficient is 0.7. Calculate for a mass flow of 1 kg/s and symmetrical blading.

(a) The blade inlet angle.

(b) The driving force on the wheel.

(c) The axial thrust.

(d) The diagram power. **[JNVU, 1990]**

Solution : Given :

$$C_1 = 900\ m/s,\ \alpha_1 = 20°,\ U = 300\ m/s\quad k = 0.7,\ \dot{m} = 1\ kg/s$$

From inlet velocity triangle

$$C_{a1} = C_1 \sin\alpha_1 = 900 \sin 20° = 307.8\ m/s$$

$$C_{w1} = C_1 \cos \alpha_1 = 900 \cos 20^\circ = 845.72 \ m/s$$

$$\therefore C_{w1} - U = 845.72 - 300 = 545.72$$

Now

$$\tan \beta_1 = \frac{C_{a1}}{C_{w1} - U} = \frac{307.8}{545.72} = 0.5640255$$

$$\therefore \beta_1 = \tan^{-1}(0.5640255) = 29^\circ.25' \qquad \text{Ans.}$$

$$V_1 = \frac{C_{a1}}{\sin \beta_1} = \frac{307.8}{\sin 29^\circ.25'} = \frac{307.8}{0.4912702} = 626.54 \text{ m/s}$$

$$V_2 = K\, V_1 = 0.7 \times 626.54 = 438.58 \ m/s$$

Given $\beta_1 = \beta_2 = 29^\circ 25'$

$$C_{w2} = V_2 \cos \beta - U = 438.58 \cos 29^\circ 25' - 300$$

$$= 382 - 300 = 82 \ m/s$$

$$\therefore \Delta C_w = C_{w1} - (-C_{w2}) = 845.72 + 82$$

$$= 927.72 \ m/s$$

(b) The driving force on the wheel

$$F = \dot{m} \Delta C_w = 1 \times 927.72 = 927.72 \ N/kg/s \qquad \text{Ans.}$$

$$C_{a2} = V_2 \sin \beta_2 = 438.58 \sin 29^\circ 25' = 215.46$$

$$\therefore \Delta C_a = C_{a1} - C_{a2} = 307.8 - 215.46 = 92.34 \ m/s$$

(c) The axial thrust $= \dot{m} \, \Delta C_a$

$$= 1 \times 92.34 = 92.34 \ N/kg/s \qquad \text{Ans.}$$

(d) The diagram power

$$= \frac{\dot{m} \, \Delta C_w \times U}{1000}$$

$$= \frac{1 \times 927.72 \times 300}{1000} = 278.316 \ kW \qquad \text{Ans.}$$

1.2 In a simple impulse turbine the nozzles are inclined at 20° to the direction of motion of the moving blades. The steam leaves the nozzle at 375 m/s. The blade speed is 165 m/s. Calculate suitable inlet and outlet angles for the blades in order that the axial thrust is zero. The relative velocity of steam as it flows over the blades is reduced by 15% by friction. Also, determine the power developed for a flow rate of 10 kg/s. **[JNVU, 1991]**

Solution : Given

$$C_1 = 375 \ m/s, \ \ \alpha_1 = 20^\circ \ \ C_{a1} = C_{a2} \text{ (No axial thrust) } \ U = 165 \ m/s$$

From inlet velocity triangle

$$C_{w1} = C_1 \cos \alpha_1 = 375 \cos 20° = 352.38 \ m/s$$

$$C_{a1} = C_1 \sin \alpha_1 = 375 \sin 20° = 128.25 \ m/s$$

$$= C_{a2}$$

$$\tan \beta_1 = \frac{C_{a1}}{C_{w1} - U} = \frac{128.25}{187.38} = 0.6844$$

$$\therefore \beta_1 = 34.38°$$ Ans.

$$V_1 = \frac{C_{a1}}{\sin \beta_1} = \frac{128.25}{\sin 34.38°} = 227.12 \ m/s$$

$$V_2 = 0.85 \ V_1 = 0.85 \times 227.12 = 193.05 \ m/s$$

$$\sin \beta_2 = \frac{C_{a2}}{V_2} = \frac{128.75}{190.05} = 0.6669$$

$$\therefore \beta_2 = 41.82°$$ Ans.

$$C_{w2} = U - V_2 \cos \beta_2$$

$$= 165 - 193.05 \ \cos 41.82°$$

$$= 165 - 143.86 = 21.14 \ m/s$$

$$\Delta C_w = C_{w1} - C_{w2} = 352.38 - 21.14 = 331.24 \ m/s$$

$$P = \frac{\dot{m} \ \Delta C_w \ U}{1000} = \frac{10 \times 331.24 \times 165}{1000} = 546.546 \ kW$$ Ans.

1.3 **The mean diameter of the blades of an impulse turbine with a single row wheel is 1 m and the speed is 3000 rpm. The nozzle angle is 18°, the ratio of linear velocity of blade to absolute velocity of steam entering moving blade is 0.42, and the relative velocity at outlet from the blades to that at inlet is 0.84. The outlet angle of the blade is to be made 3° less than the inlet angle of the blade. The steam rate of flow is 8 kg/s. Determine :**

(a) Tangential thrust on the blades.

(b) Axial thrust on the blades.

(c) Resultant thrust on the blades.

(d) Power developed. **[JNVU, 1998]**

Solution : Given :

$$D = 1m, \quad N = 3000 \ rpm \quad \alpha_1 = 18°, \quad \rho = \frac{U}{C_1} = 0.42$$

$$\frac{V_2}{V_1} = 0.84, \quad \beta_2 = \beta_1 - 3° \quad \dot{m} = 8 \ kg/s$$

$$U = \frac{\pi DN}{60} = \frac{\pi \times 1 \times 3000}{60} = 157.1 \ m/s$$

$$\frac{U}{C_1} = 0.42 \qquad \therefore C_1 = \frac{157.1}{0.42} = 374 \ m/s$$

$$C_{w1} = C_1 \cos \alpha_1 = 374 \cos 18° = 355.7 \ m/s$$

$$C_{a1} = C_1 \sin \alpha_1 = 374 \sin 18° = 115.57 \ m/s$$

$$\tan \beta_1 = \frac{C_{a1}}{C_{w1} - U} = \frac{115.57}{355.7 - 157.1} = \frac{115.57}{198.6} = 0.58$$

$$\therefore \beta_1 = \tan^{-1}(0.58) = 30.196°$$

$$\therefore \beta_2 = \beta_1 - 3° = 30.196 - 3 = 27.196°$$

$$V_1 \cos \beta_1 = C_{w1} - U = 198.6$$

$$\therefore V_1 = \frac{198.6}{\cos 30.196°} = 229.78 \ m/s$$

$$V_2 = 0.84 \times 229.78 = 193 \ m/s$$

$$C_{a2} = V_2 \sin \beta_2 = 193 \sin 27.196° = 88.2 \ m/s$$

$$V_2 \cos \beta_2 = U + C_{w2}$$

$$193 \cos 27.196° = 157.1 + C_{w2}$$

$$\therefore C_{w2} = 171.676 - 157.1 = 14.576 \ m/s$$

$$\Delta C_w = C_{w1} + C_{w2} = 355.7 + 14.576 = 370.276$$

$$\Delta C_a = C_{a1} - C_{a2} = 115.57 - 88.2 = 27.37$$

(a) $F_t = \dot{m} \, \Delta C_w$

$= 8 \times 370.276 = 2962.2 \ N$ Ans.

(b) $F_a = \dot{m} \, \Delta C_a$

$= 8 \times 27.37 = 218.96 \ N$ Ans.

(c) $F_r = \sqrt{F_t^2 + F_a^2}$

$= \sqrt{8774676.235 + 47943.48}$

$= \sqrt{8822619.717} = 2970.29 \ N$

(d) $P = \frac{\dot{m} \, \Delta C_w U}{1000} = \frac{2962.2 \times 157.1}{1000}$

$= 465.36 \ kW$ Ans.

1.4 A simple impulse steam turbine has one ring of moving blades running at 150 m/s. The absolute velocity of steam at the exit from the stage is 85 m/s at an angle 80° from the tangential direction. Blade velocity coefficient is 0.82 and the rate of steam flowing through the stage is 2.5 kg/s. If the blades are equiangular, determine (a) blade angles, (b) nozzle angle, (c) absolute velocity of steam isuing from the nozzle, (d) axial thrust, (e) Power developed. (f) If the velocity of steam reaching nozzle as 90 m/s and nozzle efficiency 0.85, find the enthalpy drop in the nozzle. **[JNVU, 2000]**

Solution : Given :

$$U = 150\ m/s,\quad C_2 = 85\ m/s,\quad \alpha_2 = 80^\circ\quad K = 0.82,\quad \dot{m} = 2.5\ kg/s\quad C_o = 90\ m/s$$

From the outlet velocity triangle

$$C_{a2} = C_2 \sin\alpha_2 = 85 \sin 80^\circ = 83.7\ m/s$$

$$C_{w2} = C_2 \cos\alpha_2 = 85 \cos 80^\circ = 14.76\ m/s$$

$$\tan\beta_2 = \frac{C_{a2}}{U + C_{w2}} = \frac{83.7}{150 + 14.76} = \frac{83.7}{164.76} = 0.5080$$

$$\therefore\ \beta_2 = \tan^{-1}(0.5080) = 26.94^\circ$$

(a) As per question, $\beta_1 = \beta_2 = 26.93^\circ$ Ans.

$$V_2 \cos\beta_2 = U + C_{w2} = 164.76$$

$$\therefore V_2 = \frac{164.76}{\cos 26.93^\circ} = 184.8\ m/s$$

$$\therefore V_1 = \frac{V_2}{K} = \frac{184.8}{0.82} = 225.3675\ m/s$$

From inlet velocity triangle

$$C_{a1} = V_1 \sin\beta_1 = 225.3675 \sin 26.93^\circ$$

$$= 225.3675 \times 0.4529 = 102.1\ m/s$$

$$C_{w1} = V_1 \cos\beta_1 + U$$

$$= 225.3675 \cos 26.93^\circ + 150$$

$$= 200.927 + 150 = 350.927\ m/s$$

$$\tan\alpha_1 = \frac{C_{a1}}{C_{w1}} = \frac{102.1}{350.927} = 0.2909$$

(b) $\therefore \alpha_1 = \tan^{-1}(0.2909) = 16.22^\circ$ Ans.

$$C_{w1} = C_1 \cos\alpha_1$$

(c) $\therefore C_1 = \dfrac{C_{w1}}{\cos\alpha_1} = \dfrac{350.927}{\cos 16.22^\circ} = 365.48\ m/s$

(d) $$F_a = \dot{m}\,(C_{a1} - C_{a2})$$

$$= 2.5\,(102.1 - 83.7) = 2.5 \times 18.4 = 46\ N$$ Ans.

(e) $$P = \frac{\dot{m}\,\Delta\,C_w\,U}{1000}$$

$$= \frac{2.5 \times (350.927 + 14.76) \times 150}{1000} = 137.13\ kW$$ Ans

(f) Applying SFEE for the nozzle

$$h_0 + \frac{C_0^2}{2000} = h_1 + \frac{C_1^2}{2000}$$

$$\therefore \frac{C_1^2 - C_0^2}{2000} = (h_0 - h_1)$$

$$\frac{(365.48)^2 - (90)^2}{2000} = \Delta\,h$$

$$\therefore \Delta\,h = 62.73\ kJ/kg$$

Considering the nozzle efficiency

$$\Delta\,h = \frac{62.73}{0.85} = 73.8\ kJ/kg$$ Ans.

1.5 Following data refer to one stage of an impulse turbine. Velocity of steam at inlet to the nozzle = 30 m/s Isentropic heat drop = 84 kJ Reheat of the steam due to friction is equal to 10% of the isentropic heat drop.

$$\frac{\text{Blade speed}}{\text{Tangential component of the steam speed}} = 0.5$$

Nozzle angle = 20°

Velocity coefficient of the blades = 0.95

Find the blade angles so that the steam may enter them without shock and leave them in an axial direction. **[JNVU, 1999]**

Solution : Given :

$$C_0 = 30\ m/s, \quad \alpha_1 = 20°, \quad \Delta\,h = 84\ kJ \quad \alpha_2 = 90° \quad K = 0.95, \quad \frac{U}{C_{w1}} = 0.5$$

10% of the isentropic heat drop = 0.9 × 84 = 75.6

For the nozzle

$$\frac{C_1^2}{2000} = 75.6 + \frac{C_0^2}{2000}$$

$$C_1 = \sqrt{75.6 \times 2000 + (30)^2}$$

$$= \sqrt{151200 + 900} = \sqrt{152100} = 390\ m/s$$

$$\frac{U}{C_{w1}} = 0.5$$

$$\therefore\ U = C_1 \cos\alpha_1 \times 0.5 = 390 \cos 20^\circ \times 0.5 = 183.24\ m/s$$

$$C_{w1} = C_1 \cos\alpha_1 = 390 \cos 20^\circ = 366.48\ m/s$$

$$C_{a1} = C_1 \sin\alpha_1 = 390 \sin 20^\circ = 133.38\ m/s$$

$$C_{w1} - U = 366.48 - 183.24 = 183.24\ m/s$$

$$\tan\beta_1 = \frac{C_{a1}}{C_{w1} - U} = \frac{133.38}{183.24} = 0.72789$$

$$\therefore\ \beta_1 = 36.05^\circ$$ Ans.

$$V_1 = \frac{C_{a1}}{\sin\beta_1} = \frac{133.38}{\sin 36.05^\circ} = 226.643\ m/s$$

$$V_2 = k\,V_1 = 0.95 \times 226.643 = 215.31\ m/s$$

From outlet velocity diagram $(C_{w2} = 0 \text{ Given})$

$$V_2 \cos\beta_2 = U$$

$$\cos\beta_2 = \frac{U}{V_2} = \frac{183.24}{215.31} = 0.85$$

$$\therefore\ \beta_2 = 31.67^\circ$$ Ans.

1.6 A single row, impulse turbine has blades whose inlet angle is 40° and exit angle 37°. The mean blade speed is 230 m/s and the nozzles are inclined at an angle of 27° to the plane of rotation of the blades. There is a 10% loss of relative velocity due to friction in the blades. The turbine uses 550 kg/hr of steam. Determine

(a) The nozzle velocity of the steam.

(b) The absolute velocity of the steam at exit.

(c) The power output of the turbine.

(d) The end thrust on the turbine. **[JNVU, 2005]**

Solution : Given :

$$\alpha_1 = 27^\circ,\quad \beta_1 = 40^\circ,\quad \beta_2 = 37^\circ,\quad U = 230\ m/s$$

From inlet velocity triangle

Using sine rule

$$\frac{U}{\sin 13^\circ} = \frac{C_1}{\sin 140^\circ} = \frac{V_1}{\sin 27^\circ}$$

$$\therefore \quad C_1 = \frac{U \sin 140^\circ}{\sin 13^\circ} = \frac{230 \times 0.64278}{0.22495} = 657.21 \ m/s$$ Ans.

$$V_1 = \frac{U \sin 27^\circ}{\sin 13^\circ} = \frac{230 \times 0.4539}{0.22495} = 464.18 \ m/s$$

$$\therefore \ V_2 = 0.9 \ V_1 = 417.762 \ m/s$$

$$C_{a1} = V_1 \sin \beta_1 = 464.18 \sin 40^\circ$$

$$C_{w1} = C_1 \cos 27^\circ = 657.21 \cos 20^\circ = 585.578 \ m/s$$

From outlet velocity triangle

$$C_{w2} = V_2 \cos \beta_2 - U$$

$$= 417.762 \cos 37^\circ - 230$$

$$= 333.639 - 230$$

$$= 103.639 \ m/s$$

$$C_{a2} = V_2 \sin \beta_2 = 417.762 \sin 37^\circ = 251.415 \ m/s$$

From the triangle

$$C_2^2 = (C_{a2})^2 + (C_{w2})^2$$

$$= (251.415)^2 + (103.639)^2$$

$$= 73950.7696$$

$$\therefore \quad C_2 = 271.93 \ m/s$$ Ans.

$$\Delta C_w = C_{w1} + C_{w2} = 585.578 + 103.639 = 689.217$$

$$P = \frac{\dot{m} \ \Delta C_w \ U}{1000} = \frac{550}{60} \times \frac{689.217 \times 230}{1000}$$

$$= 24.218 \ kW$$ Ans.

Axial thrust

$$F_a = \dot{m} \ (C_{a1} - C_{a2})$$

$$= \frac{550}{3600} (298.369 - 251.415)$$

$$= 0.153 \times 46.99 = 7.1895 \ N$$ Ans.

1.7 Steam at 7 bar and 300° expands to 3 bar in an impulse stage. The nozzle angle is 20°, the rotor blades have equal inlet and outlet angles and the stage operates with the optimum blade speed ratio. Assuming that the isentropic efficiency of the nozzle is 0.9 and that the velocity at entry to the stage is negligible, deduce the blade angle used and the mass flow required for this stage to produce 75 kW. **[JNVU, 1996]**

Solution :

From Mollier chart or use steam tables, the enthalpy drop is approximately 200 kJ/kg

$$\therefore C_1 = 44.72\sqrt{k\,\Delta h}$$

$$= 44.72\sqrt{0.9 \times 200} = 599.98 \text{ m/s}$$

$$\rho = \frac{U}{C_1} = \frac{\cos \alpha_1}{2} = \frac{\cos 20°}{2} = 0.4698$$

$$\therefore U = 0.4698\ C_1 = 0.4698 \times 599.98 = 281.87\ m/s$$

From inlet velocity triangle

$$C_{w1} = C_1 \cos 20° = 599.98 \cos 20° = 563.796\ m/s$$

$$C_{a1} = C_1 \sin 20° = 599.98 \sin 20° = 205.20\ m/s$$

$$\therefore C_{w1} - U = 563.796 - 281.87 = 281.926\ m/s$$

$$\tan \beta_1 = \frac{C_{a1}}{C_{w1} - U} = \frac{205.2}{281.926} = 0.7278$$

$$\therefore \beta_1 = 36° = \beta_2$$ Ans.

$$V_1 = \frac{C_{a1}}{\sin \beta_1} = \frac{205.2}{\sin 36°} = 349.1\ m/s$$

$$V_1 = V_2 = 349.1\ m/s$$

From outlet velocity triangle

$$C_{w2} = V_2 \cos \beta_2 - U$$

$$= 349.1 \cos 36° - 281.87$$

$$= 282.427 - 281.87 = 0.5578$$

$$\Delta C_w = C_{w1} + C_{w2} = 563.796 + 0.5578$$

$$= 564.3538\ m/s$$

$$P = \frac{\dot{m}\,\Delta C_w\,U}{1000}$$

$$\therefore \dot{m} = \frac{P \times 1000}{\Delta C_w\,U} = \frac{75 \times 1000}{564.3538 \times 281.87} = 0.47147\ kg/s$$ Ans.

1.8 **In a single stage impulse turbine the nozzle discharge the fluid on the the blades at an angle of 25° to the plane of rotation and the fluid leaves the blades with an absolute velocity of 300 m/s at an angle of 120° to the direction of motion of the blades. If the blades have equal inlet and outlet angles and there is no axial thrust, estimate the blade angle, power produced per kg/s of fluid.**

[JNVU 1996, 1997, 1999, 2000]

Solution : Given,

$$\alpha_1 = 25^\circ \qquad C_2 = 300\ m/s, \qquad \alpha_2 = 180 - 120 = 60^\circ$$

$$C_{a1} = C_{a2} \text{ (no axial thrust)}$$

From outlet velocity triangle

$$C_{a2} = C_2 \sin\alpha_2 = 300 \sin 60^\circ = 259.8\ m/s$$

$$C_{w2} = C_2 \cos\alpha_2 = 300 \cos 60^\circ = 150\ m/s$$

From the given condition

$$C_{a1} = C_{a2} = 259.8 = C_1 \sin\alpha_1$$

$$\therefore\ C_1 = \frac{259.8}{\sin 25^\circ} = 614.74\ m/s$$

$$C_{w1} = C_1 \cos\alpha_1 = 614.74\ \cos 25^\circ = 557.142\ m/s$$

$$\Delta C_w = C_{w1} + C_{w2} = 557.142 + 150 = 707.142\ m/s$$

As per question

$$\beta_1 = \beta_2$$

$$\therefore\ \tan\beta_1 = \tan\beta_2$$

$$\frac{259.8}{557.142 - U} = \frac{259.8}{150 + U}$$

or, $557.142 - U = 150 + U$

or, $2U = 557.142 - 150 = 407.142$

$$\therefore\ U = 203.571\ m/s$$

$$\therefore\ \tan\beta_1 = \frac{259.8}{557.142 - 203.571} = \frac{259.8}{353.67} = 0.734579$$

$$\beta_1 = 36.3^\circ = \beta_2$$ Ans.

$$P = \frac{\Delta C_w\, U}{1000} = \frac{707.142 \times 203.571}{1000} = 143.95\ kW/kg/s$$ Ans.

1.9 The steam reaches the nozzles of one stage of an impulse turbine at a pressure of 7 bar with negligible velocity and expands in the nozzle giving a heat drop of 300 kJ. The nozzle angle is 18° and the nozzle efficiency is 90%. The moving blade having exit angle, 25°, run at a speed of 300 m/s, when the rate of steam flow is 4.5 kg/min, the power developed at the blading is 18 kW. Calculate :

(a) The total axial thrust on the blading.

(b) The percentage friction loss in the blade channels with respect to kinetic energy corresponding to relative velocity at inlet to the blades. **[JNVU, 2004]**

Solution : Given,

$p_0 = 7\ bar, \quad \eta_n = 0.9, \quad \Delta h = 300\ kJ, \quad U = 300\ m/s$

$\alpha_1 = 18^\circ, \quad P = 20\ kW \quad \dot{m} = 4.5\ kg/\min \quad = 0.075\ kg/s$

$C_1 = 44.72\sqrt{\eta_n \times \Delta h} = 44.72\sqrt{0.9 \times 300} = 734.82$ m/s

$C_{a1} = C_1 \sin\alpha_1 = 734.82 \sin 18^\circ = 227\ m/s$

$C_{w1} = C_1 \cos\alpha_1 = 734.82 \cos 18^\circ = 698.85\ m/s$

$P = \dot{m}\ \Delta C_w\ U$

$18 \times 1000 = 0.075 \times \Delta C_w \times 300$

$\Delta C_w = 800\ m/s$

$\therefore C_{w2} = \Delta C_w - C_{w1} = 800 - 698.85$

$= 101.15\ m/s$

$$\tan\beta_1 = \frac{C_{a1}}{C_{w1} - U} = \frac{227}{698.85 - 300}$$

$$= \frac{227}{398.85} = 0.569$$

$\therefore \beta_1 = 29.64^\circ$

$C_{a1} = V_1 \sin\beta_1$

$$\therefore V_1 = \frac{227}{\sin 29.64^\circ} = \frac{227}{0.4946} = 458.923$$

$$V_2 = \frac{U + C_{w2}}{\cos\beta_2} = \frac{300 + 101.15}{\cos 25^\circ}$$

$$= \frac{401.15}{0.9063} = 442.62\ m/s$$

$C_{a2} = V_2 \sin 25^\circ = 442.62 \times 0.4226 = 187.06\ m/s$

$\therefore \Delta C_a = C_{a1} - C_{a2} = 227 - 187.06 = 39.94\ m/s$

(a) $F_a = \dot{m}\ \Delta C_a = 0.075 \times 39.94 = 2.995\ N$ Ans.

(b) Friction losses in the blades $= \dfrac{V_1^2 - V_2^2}{2}$

Kinetic energy corresponding to inlet relative velocity $= \dfrac{V_1^2}{2}$

Let × be the percentage

As per question

$$x \frac{V_1^2}{2} = \frac{V_1^2 - V_2^2}{2}$$

$$V_2^2 = (1 - x) V_1^2$$

$$(442.62)^2 = (1 - x)(458.923)^2$$

$$1 - x = 0.9302$$

$$\therefore x = 1 - 0.9302 = 0.06978$$ Ans.

1.10 In a single stage impulse turbine the mean diameter of the blade ring is one metre and the rotational speed is 3000 rpm. The steam is issued from a nozzle at 300 m/s and the nozzle angle is 20°. The blades are equiangular. If the friction loss in the blade channels is 19% of the kinetic energy corresponding to the relative velocity at the inlet to the blades, what is the power developed in the blading when the axial thrust on the blades is 100 N. **[JNVU, 2010]**

Solution : Given

$$D = 1\ m, \quad N = 3000\ rpm \quad C_1 = 300 m/s, \quad \alpha_1 = 20°, \quad F_a = 100\ N$$

Friction loss in the blade channels $= 0.19 \frac{V_1^2}{2}$

As per question

$$\frac{V_1^2 - V_2^2}{2} = 0.19 \frac{V_1^2}{2}$$

$$0.81 V_1^2 = V_2^2$$

$$\therefore \frac{V_2}{V_1} = \sqrt{0.81} = 0.9$$

$$U = \frac{\pi D N}{60} = \frac{\pi \times 1 \times 3000}{60} = 157.08\ m/s$$

$$C_{w1} = C_1 \cos \alpha_1 = 300 \cos 20° = 281.91\ m/s$$

$$C_{a1} = C_1 \sin \alpha_1 = 300 \sin 20° = 102.61\ m/s$$

$$C_{w1} - U = 281.91 - 157.08 = 124.83\ m/s$$

$$\tan \beta_1 = \frac{C_{a1}}{C_{w1} - U} = \frac{102.61}{124.83} = 0.822$$

$$\therefore \beta_1 = 39.42° = \beta_2$$

$$V_1 = \frac{C_{a1}}{\sin \beta_1} = \frac{102.61}{\sin 39.42°} = 161.59\ m/s$$

$$V_2 = 0.9\ V_1 = 0.9 \times 161.59 = 145.43\ m/s$$

$$C_{a2} = V_2 \sin \beta_2 = 145.43 \sin 39.42^\circ = 92.35 \ m/s$$

$$\Delta C_a = 102.61 - 92.35 = 10.26$$

$$F_a = \dot{m} \, \Delta C_a$$

$$100 = \dot{m} \times 10.92$$

$$\therefore \dot{m} = 9.7465 \ kg/s$$

From outlet velocity triangle

$$C_{w2} = U - V_2 \cos \beta_2$$

$$\therefore C_{w2} = 157.08 - 145.43 \cos 39.42^\circ$$
$$= 157.08 - 112.346 = 44.733 \ m/s$$

$$\Delta C_w = C_{w1} - C_{w2} = 281.91 - 44.733 = 237.177$$

$$P = \frac{\dot{m} \, \Delta C_w \, U}{1000} = \frac{9.7465 \times 237.177 \times 157.08}{1000} = 363.113 \ kW$$ Ans.

1.11 A Pparsons impulse-reaction turbine running at 400 rpm with half degree reaction develops 5 MW and uses 5.45 kg/kWhr of steam. The exit angle of the blade at exit is 1.35 times the mean blade speed. At a particular stage in the expansion the pressure is 1.7 bar and corresponding dryness fraction of the steam is 0.95. Calculate for this stage a suitable blade height, assuming that rotor diameter = 12 x Blade height. Also calculate blade velocity, the steam velocity delivers from the fixed blades and the rotor diameter. **[JNVU, 1994, 2000, 2008, 2010]**

Solution : Given

$$N = 400 \ rpm, \ C_1 = V_2 = 1.35 \ U, \ P = 5 \ MW$$

$$\dot{m} = 5.45 \ kg/kWhr \quad \alpha_1 = \beta_2 + 20^\circ \quad D = 12 \ h$$

Using steam tables, at 1.7 bar, $v_g = 1.031$ m/kg

$$\therefore v = x \, v_g = 0.95 \times 1.031 = 0.97945 \ m^3/kg$$

$$\dot{m} = \frac{5.45 \times 5000}{3600} = 7.57 \ kg/s$$

$$C_{a1} = C_1 \sin 20^\circ = 1.35 \ U \sin 20^\circ = 0.46 \ U$$

$$\therefore \dot{m} \, v = \pi D \, h \, C_{a1}$$

$$7.57 \times 0.97945 = \pi \ 12 \ h^2 \times 0.46 \ U$$

$$7.414 = 17.4 \ h^2 \ U$$

$$\therefore h^2 \ U = 0.426$$

$$U = \frac{\pi \, D \, N}{60} = \frac{\pi \times 12 \, h \times 400}{60} = 251.327 \ h \qquad (2)$$

Substituting the value of U in equation (1)

$$h^2 \times 251.327\, h = 0.426$$

$$h^3 = \frac{0.426}{251.327} = 1.69 \times 10^{-3}$$

$$h = 0.11923 \; m = 119.23 \; mm$$ Ans.

$$U = 251.327\, h = 251.327 \times 0.11923 = 30 \; m/s$$ Ans.

$$C_1 = 1.35\, U = 1.35 \times 30 = 40.5 \; m/s$$ Ans.

$$D = 12 \times 0.11923 = 1.43 \; m$$ Ans.

1.12 At a particular ring of a reaction turbine, the blade speed is 67 m/s, and the flow of steam is 4.54 kg/s, dry saturated at 1.4 bar. Both fixed and moving blades have inlet and exit angles of 35° and 20° respectively. Determine:

(a) The power developed by the pair of rings;

(b) The required blade height which is to be one tenth of the blade ring diameter. **[JNVU 2000, 2001, 2008]**

Solution : Given

$$U = 67 \; m/s, \;\; \alpha_1 = \beta_2 = 20°, \;\; \dot{m} = 454 \; kg/g, \;\; \alpha_2 = \beta_1 = 35° \qquad h = \frac{D}{10}, \; \eta_s = 0.8$$

Using sine rule in the inlet velocity triangle

$$\frac{V_1}{\sin 20°} = \frac{C_1}{\sin(180 - 35)} = \frac{U}{\sin 15°}$$

$$\therefore C_1 = U \frac{\sin 145°}{\sin 15°} = \frac{67 \times 0.573}{0.2588} = 148.48 \; m/s$$

$$V_1 = U \frac{\sin 20°}{\sin 15°} = \frac{67 \times 0.342}{0.2588} = 88.54 \; m/s$$

$$C_{a1} = C_1 \sin \alpha_1 = 148.48 \sin 20° = 50.783 \; m/s$$

$$\Delta C_w = 2\, C_1 \cos \alpha_1 - U$$

$$= 2 \times 148.48 \cos 20° - 67$$

$$= 279.051 - 67 = 212.051 \; m/s$$

$$P = \frac{\dot{m}\, \Delta C_w\, U}{1000} = \frac{4.54 \times 212.051 \times 67}{1000} = 64.5 \; kW$$ Ans.

Using steam tables

At 1.4 bar, $x_1 = 1 \quad \therefore v_{g1} = 1.237$ m³/kg

$$\dot{m} = \frac{A\, C_{a1}}{v_1} = \frac{\pi\, D\, h\, C_{a1}}{v_1} = \frac{\pi\,(10\, h)\, h \times 650.783}{1.237}$$

$$4.54 = 1289.733\ h^2$$

$$h^2 = \frac{4.54}{1289.733} = 3.52 \times 10^{-3}$$

$$h = 0.0593\ m \quad = 59.3\ mm \quad = 60\ mm$$

$$D = 10\ h \quad = 600\ mm$$ Ans.

1.13 **In a reaction turbine, the whirl velocity at inlet is 260 m/s and the mean blade speed is 150 m/s. The specific volume of steam at fixed blade outlet is 4.38 m³/kg and at moving blade outlet 4.8 m³/kg. The outlet angles of fixed and moving blades are 20° and 36° respectively. Assume carry over factor as 0.86 and efficiency of the nozzle is 90%. Determine:**

(a) Inlet,

(b) Total enthalpy drop in a pair of blade rings angles,

(c) Degree of reaction. **[JNVU, 2000, 2005, 2009]**

Solution : Given

$$\alpha_1 = 20^\circ,\ \beta_2 = 36^\circ,\ C_{w1} = 260\ m/s,\ U = 150\ m/s$$

$$V_1 = 4.38\ m^3/kg,\ V_2 = 4.8\ m^3/kg,\ \phi = 0.86,\ \eta_n = 90\%$$

From the inlet velocity triangle

$$C_{w1} = C_1 \cos\alpha_1$$

$$260 = C_1 \cos 20^\circ$$

$$\therefore C_1 = \frac{260}{\cos 20^\circ} = 276.68\ m/s$$

$$\therefore C_{w1} - U = 260 - 150 = 110\ m/s$$

$$C_{a1} = C_1 \sin 20^\circ = 276.68 \sin 20^\circ = 94.63\ m/s$$

$$\tan\beta_1 = \frac{C_{a1}}{C_{w1} - U} = \frac{94.63}{110} = 0.86$$

$$\beta_1 = \tan^{-1}(0.86) = 40.7$$ Ans.

$$C_{a1} = V_1 \sin\beta_1$$

$$94.63 = V_1 \sin 40.7^\circ$$

$$\therefore V_1 = 145.1\ m/s$$

Using continuity equation and area is constant i.e. $A_1 = A_2$

$$\dot{m}_1 = \dot{m}_2$$

$$\frac{A_1\, C_{a1}}{v_1} = \frac{A_2\, C_{a2}}{v_2}$$

$$\therefore \frac{C_{a1}}{v_1} = \frac{C_{a2}}{v_2}$$

$$\therefore C_{a2} = \frac{94.63}{4.38} \times 4.8 = 103.7 \ m/s$$

From the outlet velocity triangle

$$C_{a2} = V_2 \sin \beta_2$$

$$103.7 = V_2 \ \sin 36^\circ$$

$$V_2 = 176.43 \ m/s$$

$$C_{w2} = V_2 \cos \beta_2 - U$$

$$= 176.43 \ \cos 36^\circ - 150$$

$$= 142.736 - 150 = -7.26$$

$$\tan(180 - \alpha_2) = \frac{C_{a2}}{C_{w2}} = \frac{103.7}{7.26} = 14.276$$

$$180 - \alpha_2 = \tan^{-1}(14.276) = 85.99 = 86^\circ$$

$$\alpha_2 = 180 - 86 = 94^\circ$$ Ans.

$$C_2 \sin(180 - \alpha_2) = C_{a2}$$

$$C_2 = \frac{103.7}{\sin 94^\circ} = 103.95 \ m/s$$

Enthalpy drop in fixed blades

$$= \frac{C_1^2 - \phi\, C_2^2}{2 \times 1000 \times \eta_n} = \frac{76551.82 - 0.86 \times 10806.27}{2000 \times 0.9}$$

$$= \frac{76551.82 - 9293.39}{2000 \times 0.9}$$

$$= \frac{67258.427}{2000 \times 0.9} = 37.365 \ kJ/kg$$ Ans.

Enthalpy drop in moving blades

$$= \frac{V_2^2 - \phi\, V_1^2}{2 \times 1000 \times \eta_n} = \frac{31127.54 - 18106.44}{2000 \times 0.9}$$

$$= \frac{13021.09}{2000 \times 0.9} = 7.234 \ kJ/kg$$ Ans.

$$\Delta h = Total \ enthalpy \ drop = 37.365 + 7.234 = 44.6 \ kJ/kg$$ Ans.

$$\text{Degree of reaction} = \frac{\text{Enthalpy drop in moving blades}}{\text{Total enthalpy drop}}$$

$$= \frac{7.234}{44.6} = 16.22\%$$ Ans.

1.14 The measured consumption of a Parsons reaction turbine is 5 kg/s. The speed of turbine is 400 rpm. At a certain pair in the turbine the pressure is found to be 2 bar and dryness fraction is 0.96. The power developed by the turbine is 4.4 kW. The discharge blade tip angle is 20° for both fixed and moving blades and the axial velocity of flow is 0.72 of the blade velocity. The tip leakage steam is found to be 5%. Neglect blade thickness, determine drum diameter and blade height of the pair considered. **[JNVU 1999, 2002, 2007]**

Solution : Given

$$\dot{m} = 5\ kg/s,\ N = 400\ rpm,\ \alpha_1 = \beta_2 = 20°,$$

$$C_{a1} = 0.72\ U,\ \ P = 4.4\ kW \quad tip\ leakage = 10\%$$

D is the mean diameter = d + h

$$C_{a1} = C_1 \sin \alpha_1 = 0.72\ U$$

$$\therefore\ C_1 = \frac{0.72\ U}{\sin 20°} = 2.105\ U$$

$$U = \frac{\pi\ D\ N}{60}$$

$$= \frac{\pi\ D \times 400}{60} = 20.94\ D$$

$$\therefore C_1 = 2.105 \times 20.94\ D$$

$$= 44.087\ D$$

5% is tip leakage $\therefore$ Actual mass flow is $0.95 \times 5 = 4.75$

$$\Delta\ C_w = 2\ C_1 \cos \alpha_1 - U = 2 \times 44.087\ D \cos 20° - 20.94\ D = 61.916\ D$$

$$P = \frac{0.95\ \dot{m}\ \Delta\ C_w\ U}{1000}$$

$$4.4 = \frac{0.95 \times 5 \times 61.916\ D \times 20.94}{1000}$$

$$\frac{4.4 \times 1000}{0.95 \times 5 \times 20.94} = 61.916\ D^2$$

$$D^2 = \frac{44.236}{61.916} = 0.7144$$

$$\therefore D = 0.8452\ m = 845.2\ mm$$

$$\therefore U = 20.94\ D = 17.698$$

$$\Delta C_w = 61.916\ D = 61.916 \times 0.8452 = 52.33\ m/s$$

$$C_{a1} = 0.72\ U = 0.72 \times 17.698 = 12.743\ m/s$$

Using steam tables

$$At\ 2\ bar,\ \ v_g = 0.886\ m^3/kg$$

$$\therefore v = 0.96 \times 0.886 = 0.85056\ m^2/kg$$

$$Actual\ \dot{m} = \frac{\pi\ D\ h\ C_a}{v}$$

$$4.75 = \frac{\pi \times 0.8452 \times h \times 12.743}{0.85056}$$

$$h = 0.1194\ m$$

$$= 119.4\ mm$$ Ans.

$$\therefore\ d + h = 845.2 \qquad \therefore\ d = 725.8\ mm$$ Ans.

1.15 A stage of a steam turbine with Parson's blading delivers dry saturated steam at 2.7 bar from the fixed blades at 90 m/s. The mean blade height is 40 mm, and the moving blade exit angle is 20°. The axial velocity of the steam is three quarters of the blade velocity at the mean radius. Steam is supplied to the stage at the rate of 2.5 kg/s. The effect of the blade tip thickness on the annular area can be neglected. Calculate :

(a) the rotational speed of the wheel,

(b) the diagram power, and

(c) the enthalpy drop of the steam in this stage.

[JNVU, 1987, 2000, 2004, 2007, 2009]

Solution : Given

$$C_{a1} = 0.75\ U,\ C_1 = 90\ m/s = V_2,\ \ \beta_2 = \alpha_1 = 20°,\ \dot{m} = 2.5\ kg/s,\ \ h = 0.04$$

$$C_{a1} = C_1 \sin\ \alpha_1 = 90 \sin 20° = 30.78\ m/s$$

$$\therefore\ \ U = \frac{30.78}{0.75} = 41.04\ m/s$$

Using steam stables, at 2.7 bar

$$v_g = 0.6686\ m^3/kg = v$$

$$\dot{m} = \frac{\pi\ D\ h\ C_a}{v}$$

$$2.5 = \frac{\pi\ D \times 0.04 \times 30.78}{0.6686} \qquad \therefore\ D = 0.432\ m$$

$$U = \frac{\pi\, D\, N}{60}$$

$$N = \frac{60\, U}{\pi\, D} = \frac{60 \times 41.04}{\pi \times 0.432} = 1813.76\ rpm \qquad \text{Ans.}$$

$$\Delta\, C_w = 2\, C_1\, \cos\, \alpha_1\, - U$$

$$= 2 \times 90\, \cos\, 20^\circ - 41.04$$

$$= 169.14 - 41.04\, = 128.1\ m/s$$

$$P = \frac{\dot{m}\ \Delta\, C_w\, U}{1000} = \frac{2.5 \times 128.1 \times 41.04}{1000} = 13.14\ kW \qquad \text{Ans.}$$

Using cosine formula in the inlet velocity triangle

$$V_1^2 = C^2 + U^2 - 2V\ C_{a1}\ \cos\, \alpha_1$$

$$= 90^2 + 41.04^2 - 2 \times 41.04 \times 90\, \cos\, 20^\circ$$

$$= 8100 + 1684.28 - 6941.697$$

$$= 9784.28 - 6941.697$$

$$= 2842.58$$

$$\therefore\ V_1 = 53.315\ m/s$$

Enthalpy drop in the moving blades

$$= \frac{V_2^2 - V_1^2}{2 \times 1000} = \frac{C_1^2 - V_1^2}{2000}$$

$$= \frac{90^2 - 53.315^2}{2000}$$

$$= \frac{8100 - 2842.58}{2000} = \frac{5257.42}{2000}$$

$$= 2.628\ kJ/kg$$

Therefore,

Total enthalpy drop per stage = $2 \times 2.628\ = 5.257\ \ kJ/kg$ Ans.

EXERCISES

A. Theory

1.1 Differentiate between steam turbine and steam engine. What are the advantages of steam turbine over steam engine.

1.2 Differentiate between impulse and reaction turbines

1.3 Give the complete classification of steam turbines

1.4 Describe the working of impulse turbine. Sketch pressure and velocity variations along the axis of the turbine for a single stage turbine.

1.5 Describe the working of impulse reaction turbine. Sketch pressure and velocity variations along the axis of the turbine for a single stage turbine. Draw also h-s chart.

1.6 Draw neat sketches of impulse and impulse reaction turbine blading and explain.

1.7 What is compounding? Discribe various methods of compounding.

1.8 Sketch and explain pressure compounded impulse turbine showing the pressure and velocity variations along the axis. What are its merits and demerits?

1.9 Sketch and explain velocity compounded impulse turbine, showing the pressure and velocity variations along the axis. What are its merits and demerits.

1.10 Sketch and explain pressure velocity compounded impule turbine, showing the pressure and velocity variations along the axis. Why this type of arrangement is rarely used.

1.11 Describe various ways of compounding reaction turbine.

Section B : Unsolved problems

1.1 A single row, impulse turbine has a mean blade speed of 215 m/s. Nozzle entry angle is at 30° to the plane of rotation of the blades. The steam velocity from the nozzles is 550 m/s. There is a 15% loss of relative velocity due to friction across the blades. The absolute velocity at exit is along the axis of the turbine. The steam flow through the turbine is at the rate of 700 kg/hr. Determine (a) inlet and exit angle of the blades, (b) the absolute velocity of the steam at exit, (c) the power output of the turbine. [(a) 46.46°, 48.18°, (b) 243 m/s, (c) 19.91 kW].

1.2 Steam issues from the nozzles, inclined at an angle of 22° to the direction of motion, with a velocity of 680 m/s on the blades of a single stage impulse turbine. The diameter of the blade ring is 1.25 m. The blade angles at inlet and exit are the same and equal to 36°. Assuming that the steam enters the blade without shock and that the flow over the blades is frictionless, determine, (a) speed of the turbine rotor in rpm (b) absolute velocity of the steam leaving the blades (c) the torque on the turbine rotor wen the flow of steam is 2500 kg/hr [4276 rpm, 264.37 m/s, 304.345 Nm].

1.3 Steam issues from the nozzles of a single stage impulse turbine at 1000 m/s and the nozzles are inclined at 24° to the direction of motion of the blades which have a speed of 400 m/s. The blade angles at inlet and exit are equal. If the steam enters the blades without shock and the flow over the blades is frictionless, determine the inlet angle of the blades. Determine, also the force exerted on the blades in the direction of their motion and the power developed when the flow of steam is 4000 kg/hr. [38.379°, 1135 N, 454 kW]

1.4 In a single stage impulse turbine, the steam velocity at nozzle outlet is 300 m/s, the nozzle angle is 18% and the mean blade speed is 144 m/s. Assume that the outlet angle of blades is 3° less than inlet angle and that the relative velocity of steam at outlet from the blade is 0.84 of the relative velocity at entrance.

If the power developed is 1000 kW. Calculate the mass of steam that must pass through the turbine per second. Neglect disc friction and leakage losses [26.33 kg/s].

1.5 A single stage impulse turbine rotor has a diameter of 1.2 m running at 3000 rpm. The nozzle angle is 18°, Blade speed ratio 0.42. The ratio of the relative velocity at outlet to the relative. Velocity at inlet is 0.9. The outlet angle of the blade is 3° smaller than the inlet angle. The steam flow rate is 5 kg/s. Find the following (a) velocity of whirl, (b) Axial thrust on end bearing, (c) Blade angle, (d) Power developed [$C_w = C_{w1} + C_{w2} = 426.833 + 32.248 = 459.081$ Fa = 126.35 N, $\beta_1 = 30.194°$, $\beta_2 = 27.194°$, $432.23\ kW$].

1.6 In a reaction stage of a steam turbine the nozzle angle is 20° and the absolute velocity of the steam at inlet to the moving blade is 240 m/s. The blade velocity is 210 m/s. If the blading is designed for 50% reaction, determine (a) the blade angles at inlet and exit, (b) the enthalpy drop per kg of steam in the moving blades and in the complete stage, (c) The diagram power for a steam flow of 1 kg/s (d) The diagram efficiency [79°12′, 25.28 kJ/kg, 50.56 kJ/kg, 504 kW, 93.5%].

1.7 At a particular stage of a reaction steam turbine, the mean blade speed is 60 m/s and the steam is at a pressure of 350 kPa with a temperature of 175°C. Fixed and moving blades at this stage have inlet angle 30° and exit angle 20°. Determine (a) the blade height at this stage if the blade height is 1/10 the mean blade ring diameter and the steam flow is 13.5 kg/s, (b) the power developed by a pair of fixed and moving blade rings at this stage, (c) the specific enthalpy drop in kJ/kg at the stage if the stage efficiency is 85% [h = 64 mm, P = 218 kW, 19.1 kJ/kg].

1.8 The following data refers to a particular stage of a Parsons reaction steam turbine. Speed of the turbine = 1500 rpm, mean diameter of the rotor = 1 m, stage efficiency = 80%, Blade outlet angle = 20%, speed ratio = 0.7. Determine the available isentropic enthalpy drop in the stage [13 kJ].

1.9 In a reaction turbine, the blade tips are inclined at 35° and 20° in the direction of motion. The guide blades are of the same shape as the moving blades, but reversed indirection. At a certain place in the turbine, the drum diameter is one metre and the blades are 100 mm high. At this place, steam has pressure of 1.75 bar and dryness fraction of 0.935. If the speed of the turbine is 2500 rpm and the steam passes through the blades without shock, find the mass of steam flow and the power developed in the ring of the moving blades [39.4317 kg/s, 2585 kW].

1.10 The following particulars refers to a stage of a Parsons steam turbine, comprising one ring of fixed blades and one ring of moving blades. Mean diameter of blade ring = 70 cm, rpm = 3000, steam velocity at exit of blades = 160 m/s. Blade outlet angle = 20°, steam flow through blades = 7 kg/s. Draw a neat velocity diagram and find the following. (i) Blade inlet angle, (ii) Tangential force on the ring of moving blades, and (iii) Power developed in the stage [33.6°, 2635.9 N, 289.67 kW].

1.11 The outlet angle of a blde of Parsons turbine is 20° and the axial velocity of flow of steam is 0.5 times the mean blade velocity. Draw the velocity diagram for a stage consisting of one fixed and one moving row of blades. Given that mean diameter is

710 mm. Speed of rotation is 3000 rpm. Find the inlet angle of blades if the steam is to enter the blade channels without shocks. If the blade height is 64 mm, the mean steam pressure 5.6 bar, the steam dry saturated (v_2 = 0.3434 m³/kg), find the power developed in the stage [β_1 =53.2°, 5035 kW].

1.12 The blade angle of both fixed and moving blades of a reaction steam turbine are 35° at the receiving tips and 20° at the discharging tips. At a certain point in the turbine the drum diameter is 1.37 m and the blade height is 127 mm. The pressure of steam supply to a ring of a fixed blades at this point is 1.25 bar and the dryness fraction is 0.925. Find the work done in next row of moving blades for 1 kg of steam at 600 rpm, the steam passing through the blades without shock. Assuming an efficiency of 85% for the pair of rings of fixed and moving blades, find the heat drop in the pair and the state of steam at entrance to the next row of fixed blades [Work done = 7002.4 Nm, 8.95 kJ].

1.13 A Parsons turbine develops 1140 kW at 400 rpm and consumes 5.8 kg of steam per kWhr. Steam is supplied at 11 bar and 260°C and the isentropic efficiency of expansion is 85%. The blade angles are 35° and 20° at inlet and outlet tips respectively. Find the drum diameter and blade height at a stage where the pressure is 1.4 bar. A height to drum diameter ratio of 1/12 is recommended. Find the power developed at that stage [D_o = 1.8352 m, h = 0.0696 m, U = 18.95 m/s, ΔC_w = 61.77, P = 2.924 kW].

1.14 Steam at 15 bar and 350°C is expanded through a 50% reaction turbine to a pressure of 0.14 bar. The stage η is 75% for each stage, and the reheat factor is 1.04. The expansion is to be carried out in twenty stages and the diagram power is required to be 12000 kW. Calculate the flow of steam required, assuming that the stages all develop equal work. At one stage the pressure is 1 bar and the steam is dry saturated. The exit angle of the blades is 20°, and the blade speed ratio is 0.7. If the blade height is 1/12 of the blade mean diameter, calculate the value of the mean blade diameter and the rotor speed [U = 141.4 m/s, C_a = 69.1 m/s, D = 1.298 m, N = 2081 rpm].

1.15 The following particulars refer to one stage of an impulse reaction turbine: Mean diameter = 960 mm, RPM = 3000, Nozzle outlet angle = 20° Nozzle height = 120 mm. Blade height = 120 mm, specific volume of steam (a) at nozzle outlet = 4.4 m³/kg, (b) at blade outlet = 4.8 m³/kg steam velocity at nozzle outlet = 275 m/s, power developed by blades = 265 kW. Calculate the heat drop in the stage, the degree of reaction, the outlet angle of moving blades and the gross stage efficiency. Assume that the expansion efficieny is 0.94 and that the carryover coefficient is 0.81. Neglect the effect of nozzles and blade thickness [35.475 kJ/kg, 6.036 kJ/kg].

1.16 Dry saturated steam is supplied to an impulse reaction turbine at 4 bar and expands with an efficiency of 0.75. The blade velocity is 50 m/s and the axial velocity varies from 0.58 to 0.78 of blades velocity. Determine final pressure and useful heat drop if the final dryness is 0.985. Assuming exit angles of 20° determine work done per kg of steam for a pair of blades situated in the middle and estimate the total number of pairs [2.8 bar, 7.05 kJ/kg, 8.46 9 pairs].

Chapter 2

STEAM TURBINES: VARIOUS EFFICIENCIES, ENERGY LOSSES, CONSTRUCTION DETAILS AND COMPONENTS

2.1 BLADING EFFICIENCY

In an impulse turbine the energy supplied to the blades is the kinetic energy at inlet to the moving blades. No expansion takes place in the moving blades.

The ratio of work done to the energy supplied in a stage is called blading efficiency or diagram efficiency.

$$\text{Diagram efficiency, } \eta_d = \frac{\text{Rate of doing work per kg of steam}}{\text{Energy supplied per kg of steam}}$$

$$\eta_d = \frac{\text{Diagram work}}{\text{Energy supplied}}$$

$$= \frac{\Delta C_w U}{C_1^2 / 2}$$

$$= \frac{2 \Delta C_w U}{C_1^2} \quad (2.1)$$

In fact the complete stage consists of set of nozzles and moving blade row, and the stage efficiency includes the performance of nozzle also.

$$\text{Stage efficiency} = \frac{\text{Diagram work}}{\text{Energy supplied to rotor}}$$

$$\eta_s = \frac{\Delta C_w U}{(h_0 - h_1)} \quad (2.2)$$

where h_0 = enthalpy of steam at entry to the nozzle.

h_1 = enthalpy of steam at exit to the nozzle.

The maximum velocity of the steam impinging on the blades is given by

$$C_1 = \sqrt{2(h_0 - h_1)}$$

Assuming velocity at inlet to the nozzles is negligible.

The nozzle efficiency, $\eta_n = \dfrac{C_1^2}{2\,(h_0 - h_1)}$

or, $(h_0 - h_1) = \dfrac{C_1^2}{2\,\eta_n}$

Substituting in equation (2.2)

$$\eta_s = \frac{2\,\Delta\,C_2\,U}{C_1^2} \times \eta_n$$

or, $$\eta_s = \eta_d \times \eta_n \qquad (2.3)$$

Therefore,

Stage efficiency = Diagram efficiency x Nozzle efficiency

2.2 OPTIMUM OPERATING CONDITIONS

From equation (2.1) the diagram efficiency is given by

$$\eta_d = \frac{2\,\Delta\,C_w\,U}{C_1^2}$$

For the purpose of analysis, $\Delta\,C_w$ can be expressed as

$$\Delta\,C_w = V_1 \cos\beta_1 + V_2 \cos\beta_2 \qquad (2.4)$$

Also, if the blade velocity coefficient, K = 1 (i.e. $V_2 = V_1$), and if the blades are symmetrical (i.e. $\beta_2 = \beta_1$), then we have

$$\Delta\,C_w = 2\,V_1 \cos\beta_1$$

But, $V_1 \cos\beta_1 = C_1 \cos\alpha_1 - U$ (see Fig. 1.5)

Hence, $\Delta C_w = 2\,(C_1 \cos\alpha_1 - U)$

and Rate of doing work per kg/s

$$= 2\,(C_1 \cos\alpha_1 - U)\,U \qquad (2.5)$$

$$\therefore \eta_d = \frac{2\,(C_1 \cos\alpha_1 - U)\,U}{C_1^2 / 2} = \frac{4U}{C_1}\left(\cos\alpha_1 - \frac{4}{C_1}\right)$$

Let $\rho = \dfrac{U}{C_1}$ = blade speed ratio

$$\therefore \eta_d = 4\,\rho\,(\cos\alpha_1 - \rho)$$

$$= 4\left(\rho \cos\alpha_1 - \rho^2\right) \qquad (2.6)$$

From equation (2.5) the rate of doing work on the blade wheel per kg/s steam flow is given by $2\,U\,(C_1 \cos\alpha_1 - U)$. For a given velocity C_1, and a given blade velocity U, it can be seen that the rate of doing work is maximum when $\cos \alpha_1 = 1$ (i.e. when $\alpha_1 = 0$). For this value of α_1 the axial flow component would be zero. It is necessary to have an axial flow component to allow the steam to reach the blades properly and to clear the blades on leaving. As α_1 is increased the rate of working on the blades is reduced, but the blade annulus area required for a given mass flow is reduced since the axial flow component of the velocity is increased. Further, the surface area of the blades will be reduced at higher values of α_1, and this means friction losses will be less. A selection of α_1, lie between 15° and 30°. For a fixed value of α_1 the optimum blade speed ratio for maximum diagram efficiency can be obtained by differentiating equation (2.6) and putting the result equal to zero.

i.e. $$\eta_d = 4\left(\rho \cos \alpha_1 - \rho^2\right)$$

$$\therefore \frac{d\,\eta_d}{d\rho} = 4\left(\cos \alpha_1 - 2\,\rho\right) = 0$$

or, $$\rho = \frac{\cos \alpha_1}{2} \tag{2.7}$$

Substituting in equation (2.5), we get

$$\eta_{d\,\max} = 4\left(\frac{\cos^2 \alpha_1}{2} - \frac{\cos^2 \alpha_1}{4}\right)$$

$$= 4\left(\frac{\cos^2 \alpha_1}{4}\right)$$

$$= \cos^2 \alpha_1 \tag{2.8}$$

The rate of doing work corresponding to the maximum diagram efficiency is then given by substituting in equation (2.5)

i.e. $$2\left(C_1 \cos \alpha_1 - U\right)U$$

$$= 2\,U\left[2\,U - U\right]$$

$$= 2U\left[U\right]$$

$$= 2U^2 \tag{2.9}$$

i.e. maximum efficiency rate of doing work per kg/s

$$= 2\,U^2 \tag{2.10}$$

The variation in η_d with ρ is shown in Fig. 2.1.

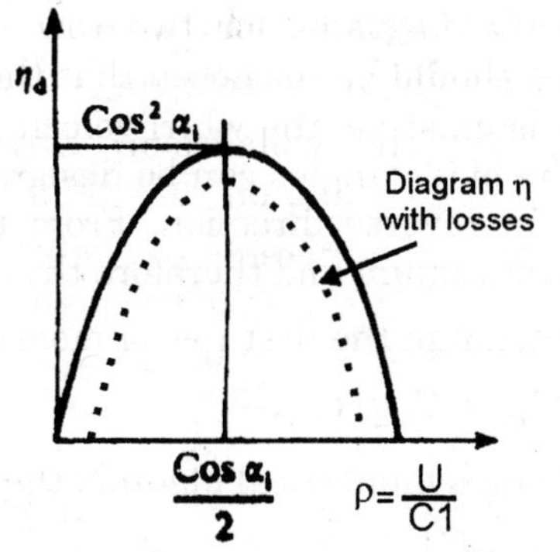

Fig. 2.1. Variation in η_d with ρ.

2.3 CALCULATIONS FOR VELOCITY COMPOUNDED IMPULSE TURBINE

Fig. 2.2 shows a two row velocity compounded stage.

The flow through the nozzles and first moving row is similar to the flow through a simple impulse stage, except that the ratio U/C_1 is much smaller. The fluid leaves the first row of rotor blades with a high velocity C_2 at an angle α_2 and enters a row of fixed guide blades having an inlet angle α_2. No further expansion takes place and these blades merely turn the stream into the direction required for entry to a second row of rotor blades, i.e. C_3 equals C_2. The guide blades are often made symmetrical, with the outlet α_3 angle equal to the inlet angle α_2, and the blade height will then be the same at inlet and outlet to give equality of areas. The geometry of the velocity triangles then fixes the magnitude V_3 and direction β_3 of the relative velocity at inlet to the second moving row. If these blades are also symmetrical, β_4 is equal to β_3 and V_4 is equal to V_3.

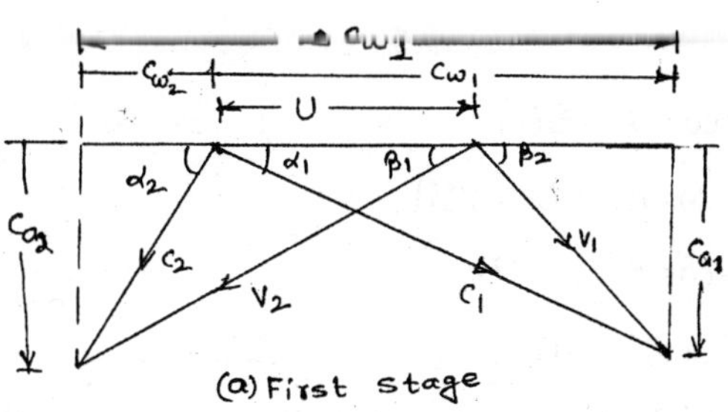

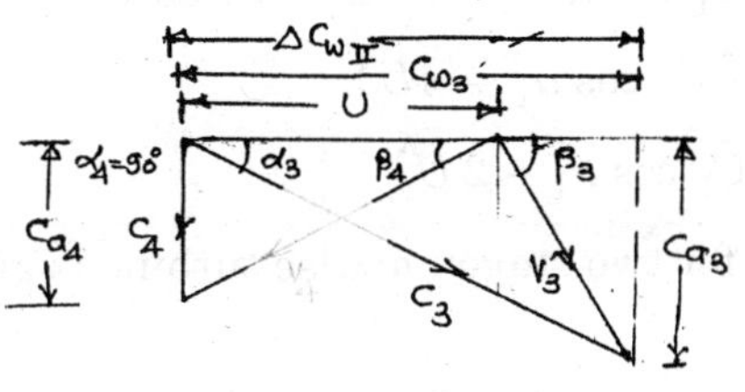

Fig. 2.2. Velocity triangles for two row velocity compounded impulse turbine - axial discharge

Fig. 2.2 shows the velocity diagrams for two row velocity compounded impulse turbines. The speed and angles should be so chosen that the final absolute velocity of the steam leaving the second row is axial i.e. the whirl velocity at exit is zero. This happens when angle is equal to 90°. The blade angles can be designed to satisfy this condition by solving the velocity diagram in reverse direction. From this arrangement the kinetic energy carried by the steam is minimum and therefore the efficiency becomes maximum.

The work done per kg of steam in the first row of moving blades is given by

$$W_1 = \Delta C_w U = (C_{w1} + C_{w2}) U$$

If there is no friction loss and symmetrical blading, then

$$W_1 = 2U (C_1 \cos \alpha_1 - U)$$

Work done per kg in the second row of moving blades is given by

$$W_2 = \Delta C_w U = (C_{w3} + C_{w4}) U$$

$$= C_{w3} . U$$

$$= (V_3 \cos \beta_3 + V_4 \cos \beta_4) U$$

Here $\beta_3 = \beta_4$ and $V_3 = V_4$

$$\therefore W_2 = 2 U V_3 \cos \beta_3$$

$$= 2U (C_3 \cos \alpha_3 - U)$$

$[C_3 = C_2 \text{ and } \alpha_3 = \alpha_2]$

$$= 2U (C_3 \cos \alpha_2 - U)$$

$$= 2U (V_2 \cos \beta_2 - U - U)$$

$$= 2U (V_2 \cos \beta_2 - 2U)$$

$[\beta_2 = \beta_1 \text{ and } V_2 = V_1]$

$$= 2U (V_1 \cos \beta_1 - 2U)$$

$$= 2U (C_1 \cos \alpha_1 - U - 2U)$$

$$= 2U (C_1 \cos \alpha_1 - 3 U)$$

Total work done $= W_1 + W_2 = W_t$

$$W_t = 2 U (C_1 \cos \alpha_1 - U) + 2 U (C_1 \cos \alpha_1 - 3 U)$$

$$= 2U (2 C_1 \cos \alpha_1 - 4U)$$

$$= 4U (C_1 \cos \alpha_1 - 2 U) \quad (2.11)$$

The diagram efficiency for two stage impulse turbine is given by

$$\eta_d = \frac{Wt}{C_1^2 / 2}$$

$$= \frac{4U(C_1 \cos \alpha_1 - 2U)}{C_1^2 / 2}$$

$$= 8U\left(\cos \alpha_1 - \frac{2U}{C_1^2}\right)$$

$$= 8\rho(\cos \alpha_1 - 2\rho) \tag{2.12}$$

where, $\rho = U / C_1$

The diagram efficiency for two stage velocity compounded impulse turbine will be maximum

When $$\frac{d\eta_d}{d\rho} = 0 = \frac{d}{d\rho}\left(8\rho \cos \alpha_1 - 16\rho^2\right)$$

or, $$8 \cos \alpha_1 - 32\rho = 0$$

or, $$\rho = \frac{\cos \alpha_1}{4} \tag{2.13}$$

Substituting the valve of ρ in equation (2.12), we get

$$(\eta_d)_{max} = \cos^2 \alpha_1 \tag{2.14}$$

The maximum work done per kg of steam is obtained by substituting the valve of ρ in equation (2.11)

$$(W)_{max} = 8U^2 \tag{2.15}$$

Thus, for any given blade speed, a two row stage can utilise four times the enthalpy drop of a simple impulse stage without reduction in diagram or blading efficiency. Thus velocities are of course high in the nozzles and first rotor row, and friction losses may be rather large.

The above analysis is done for two stages only. Similar procedure is adopted for more stages.

A trigonometrical analysis of the vector diagrams, similar to that given above will disclose the optimum $\rho = U / C_1$ of a velocity compounded turbine with n stages. Much involved arithmetical working can be avoided by assuming frictionless flow and symmetrical blading. The result of this approach reveals that the maximum diagram efficiency occurs when

$$\rho = \frac{1}{2n} \cos \alpha_1 \tag{2.16}$$

Friction aside, the magnituide of the maximum diagram efficiency remains unaffected by the number of symmetrical blade stages at a value of $\cos^2 \alpha_1$. However, the introduction of additional blade rings naturally produces increased frictional losses and the maximum diagram efficiency suffer on this account. Fig. 2.3 summarizes the effects of blade speed/steam speed ratio on the diagram efficiency for various numbers of velocity compounded stages.

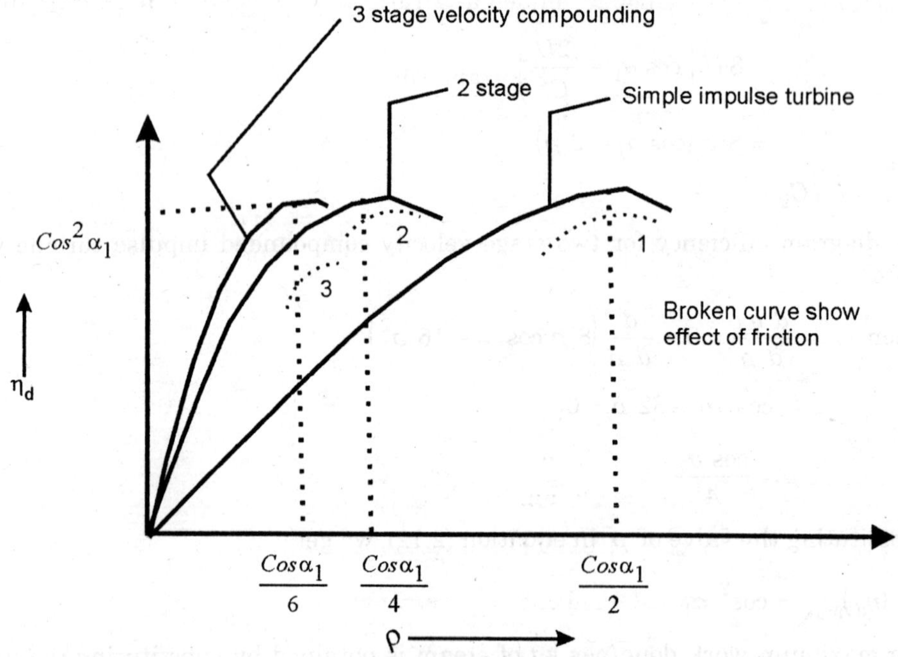

Fig. 2.3. Effect of ρ on η_d of velocity compounded turbines

The work done in the last row is given by 1/2n of total work.

The utility of the last row decreases with an increase in number of stages. In practice, more than two rows are rarely preferred.

2.3.1 Advantages and limitations

Advantages

(1) The space requirement is less

(2) The turbine housing need not to make strong because the total pressure drop occurs in the nozzle itself.

(3) Initial cost is less because fever number of stages are required.

Limitations

(1) Due to high velocity of steam friction losses are large

(2) The maximum diagram efficiency and efficiency range decreases with the increase in number of stages (Fig. 2.3).

(3) The power developed in each successive blade row decreases with increase in number of rows. Still the rows require some space, material and initial cost. Therefore all the stages are not used with equal economy.

2.4 OPTIMUM BLADE SPEED RATIO

The diagram efficiency is defined as a ratio of work done to the energy supplied in a stage.

$$\eta_d = \frac{\text{Diagram work}}{\text{Energy supplied to the moving blades}} \tag{2.17}$$

Energy supplied to the moving blades

$$= \frac{C_1^2}{2} + \left(\frac{V_2^2 - V_1^2}{2}\right)$$

$$= \frac{C_1^2}{2} + \frac{C_1^2}{2} - \frac{V_1^2}{2} \qquad [\text{As } C_2 = V_1]$$

$$= C_1^2 - \frac{V_1^2}{2}$$

From the velocity triangle (Fig. 2.3)

$$V_1^2 = C_1^2 + U^2 - 2\,U\,C_1 \cos\alpha_1$$

Therefore, energy supplied to the moving blades

$$C_1^2 - \frac{\left(C_1^2 + U^2 - 2U\,C_1 \cos\alpha_1\right)}{2}$$

$$= \frac{C_1^2 - U^2 - 2\,U\,C_1 \cos\alpha_1}{2} \tag{2.18}$$

Diagram work = $\Delta C_w\,U$

$$= (2\,C_1 \cos\alpha_1 - U)\,U \tag{2.19}$$

Substituting the values from equations (2.18) and (2.19) into equation (2.17) we get,

$$\eta_d = \frac{2\,(2\,C_1 \cos\alpha_1 - U)\,U}{\left(C_1^2 - U^2 - 2\,U\,C_1 \cos\alpha_1\right)}$$

$$\eta_d = \frac{\dfrac{2U}{C_1}\left(2\cos\alpha_1 - \dfrac{U}{C_1}\right)}{1 - \left(\dfrac{U}{C_1}\right)^2 + 2\left(\dfrac{U}{C_1}\right)\cos\alpha_1} \qquad \left(\text{As } \rho = \frac{U}{C_1}\right)$$

$$= \frac{2\,\rho\,(2\cos\alpha_1 - \rho)}{1 - \rho^2 + 2\,\rho\cos\alpha_1} \tag{2.20}$$

The equation (2.20) is rearranged in a manner shown below

$$\eta_d = \frac{2\left(1 - \rho^2 + 2\,\rho\cos\alpha_1\right) - 2}{\left(1 - \rho^2 + 2\,\rho\cos\alpha_1\right)}$$

$$= 2 - \frac{2}{\left(1 + 2\,\rho\cos\alpha_1 - \rho^2\right)}$$

The η_d becomes maximum when the value of $\left(1 + 2\,\rho\cos\alpha_1 - \rho^2\right)$ becomes maximum. Therefore,

$$\frac{d}{d\,\rho}\left(1 + 2\,\rho\cos\alpha_1 - \rho^2\right) = 0$$

or, $\quad 2\cos\alpha_1 - 2\,\rho = 0$

or, $\quad \rho = \cos\alpha_1 = \dfrac{U}{C_1} \qquad (2.21)$

Substituting the value of ρ in equation (2.20), we get

$$\left(\eta_d\right)_{max} = \frac{2\cos\alpha_1\left(2\cos\alpha_1 - \cos\alpha_1\right)}{1 - \cos^2\alpha_1 + 2\cos^2\alpha_1}$$

$$= \frac{4\cos^2\alpha_1 - 2\cos^2\alpha_1}{1 + \cos^2\alpha_1}$$

$$= \frac{2\cos^2\alpha_1}{1 + \cos^2\alpha_1} \qquad (2.22)$$

The maximum diagram work is obtained by substituting the value of ρ in equation (2.19)

$$(\text{Diagram work})_{max} = (2\,U - U)\,U$$

$$= U^2 \qquad (2.23)$$

For the optimum value of ρ a velocity diagram as shown in Fig. 2.4 is obtained.

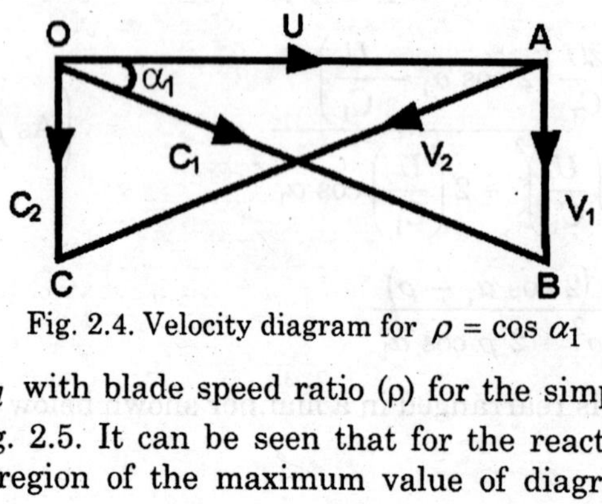

Fig. 2.4. Velocity diagram for $\rho = \cos\alpha_1$

The variation of η_d with blade speed ratio (ρ) for the simple impulse and reaction stage, are shown in Fig. 2.5. It can be seen that for the reaction turbine the curve is reasonably flat in the region of the maximum value of diagram efficiency, so that a variation in $\cos\alpha_1$, and hence $\rho = U/C_1$, can be accepted without much variation in the diagram efficiency from the maximum value.

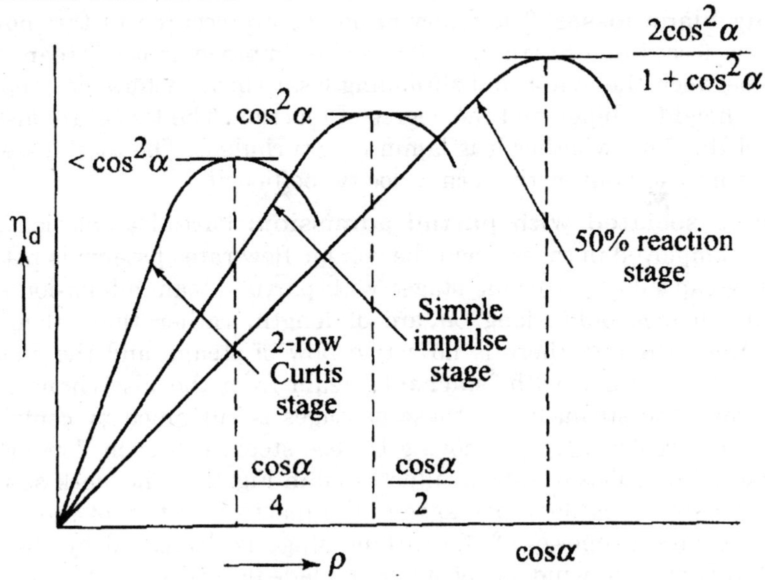

Fig. 2.5. Variation of η_d v/s ρ for simple impulse 2 row Curtis stage and 50% reaction turbine

2.5 ENERGY LOSSES IN STEAM TURBINE

The energy losses means that parts of energy which do not take part in energy conversion from one form to another. In other words, the increase in heat energy required for doing mechanical work in actual practice as compared to the theoretical value in which expansion is assumed to take place isentropically, is called as energy losses in steam turbines.

All the turbine losses grouped in two categories, namely

Internal Losses and External Losses

2.5.1 Internal Losses

In Internal losses are such losses which are directly connected with steam conditions while flowing through the steam turbines. The various losses enumerated as follows :

(1) Governor valve losses: The flow of steam through the stop or regulating or governor valves. In these valves throttling of steam takes place. Throttling is constant enthalpy process, the enthalpy drops available for work is reduced. This loss may be of the order of 5 to 10%.

(2) Nozzle loss : The nozzle efficiency is defined as the actual enthalpy drop to the isentropic enthalpy drop. Since the actual enthalpy drop is less than in the isentropic case, hen the actual velocity of the fluid leaving the nozzle is less than that obtained with isentropic expansion. The losses are caused due to steam entering to the nozzles, frictional resistance of the nozzle walls, viscous friction between particles, deflection of the flow, growth of the boundary layer, turbulence in the wake. A better aerodynamic design and a high degree of surface finish reduce the nozzle loss.

(3) Moving blade losses: The following losses are occured in the moving blades. These losses are: losses due to trailing edge wake, impingement of steam through the annular space between the stator and shrouding loss. These factors also depends on the relative velocity height of blade and the degree of reaction. The losses are also influenced by the nature of the flow, whether it is laminar or turbulent. The total losses in moving blades are taken into account by the term velocity coefficient.

(4) Losses associated with partial admission: Partial admission of steam to turbine stages is employed in cases when the volume flow rate of steam is not high, i.e. in turbines of a low capacity. In turbine stages with partial steam admission steam is fed onto the moving blades only along on arc of length, rather than along the entire circumerence. Along the arc, there is no active flow of steam, and the blade passages oppositing this arc are filled with `stagnant' steam from the disc chamber. Owing to rotation of the disc, the steam filling these passages is entrained by centrifugal forces and moves from the roots to tips of moving blades; steam can even flow over from one side of a blade onto the opposite side, as illustrated in Fig. 2.6. The work associated with this motion of the steam in blade passages of the inactive portion of the arc of moving blades, is lost, i.e. useful energy of the turbine stage is decreased by the energy loss associated with this motion (windage) of steam in blade passages.

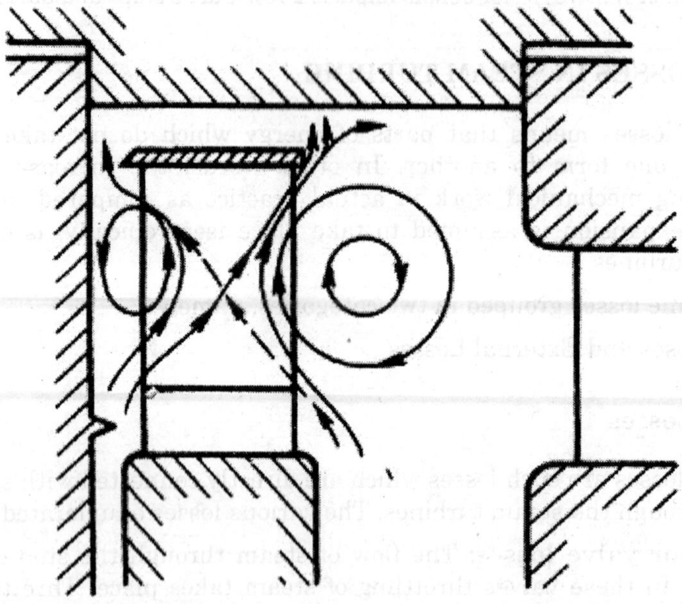

Fig. 2.6. Windage currents in a partial admission turbine stage.

The windage power can be influenced substantially by the shape and dimensions of the disc chamber in which the moving blades pass along the inactive portion of the arc. If the moving blades are protected by an antiwindage apron, as shown in Figs. 2.7 and 2.8, the coefficient decreases by 50 to 67% compared with that for an unprotected disc. In that case, the windage flow rate of steam decreases substantially, since the apron hinders steam motion.

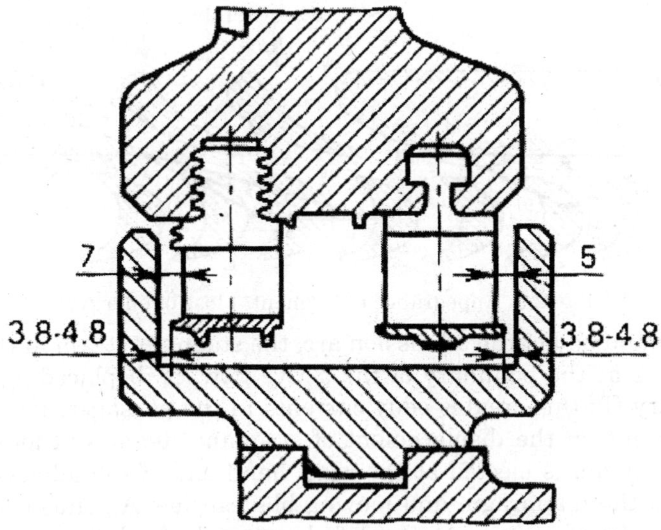

Fig. 2.7. Anti-windage apron in a row partial admission turbine stage.

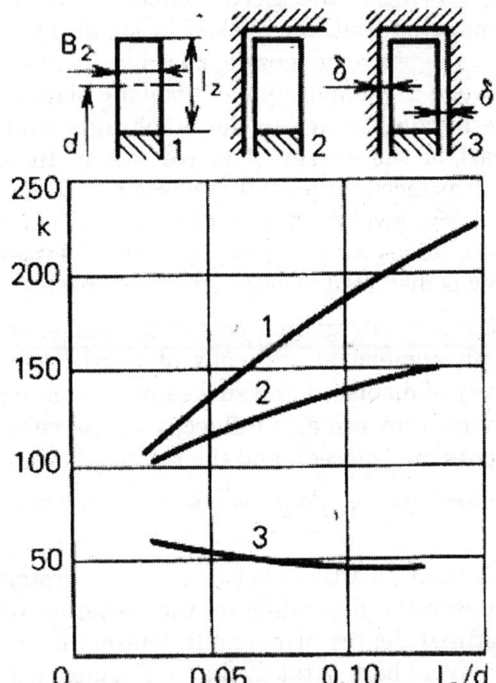

Fig. 2.8. Effect of anti-windage apron and relative blade height on windage loss.

The windage loss of energy is one of the component losses associated with partial steam admission. Another component is the loss at the ends of admission arcs (at the ends of nozzle segments). In order to disclose the nature of this loss, consider the scheme of steam flow in a partial-admission turbine stage (Fig. 2.9).

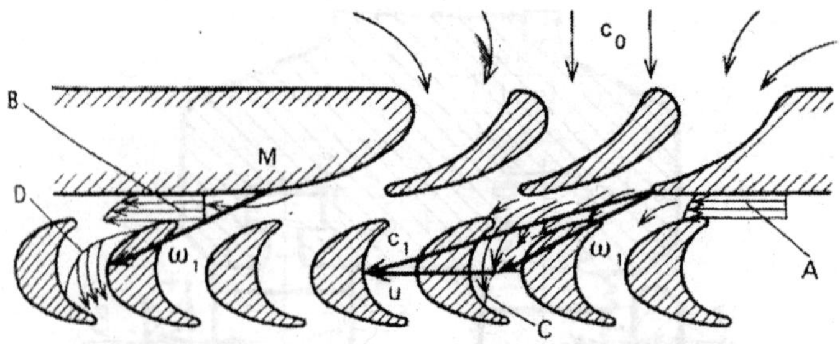

Fig. 2.9. Appearance of segmental loss of energy.

At the right hand end of the admission arc, the stagnant steam in the blade passage, as the latter moves at the boundary of the active flow, is displaced by the jet of active steam (at boundary C); this creates eddy currents in the passages. Part of the energy of active steam is spent on the displacement of stagnant steam and formation of eddies. Besides, stagnant steam is ejected at the right hand end of the admission arc from the clearance between the diaphragm and moving blades (flow A), thus consuming another portion of the energy of steam. At the left hand end of admission arc, as the blade passage moves away from the active steam jet, stagnant steam is sucked in from the clearance into the passage owing to the inertial effect of the active steam remaining in the passage, which is separated from the sucked in steam by boundary D. In addition, active steam leaks from the extreme nozzle passage at the left hand end into the meridional clearance between the diaphragm and moving blades owing to `sticking' of the steam jet of the extreme nozzle passage to the diaphragm wall (flow B). Thus, suction and leakage consume part of the energy of active steam. In order to diminish steam leakage into meridional clearance in a partial admission turbine stage, the design degree of reaction is chosen at a low level (0.02 to 0.06). In addition to the component losses mentioned above, segment losses also include the energy losses in the extreme nozzle passages in which the flow is distorted by edge effects of nozzles.

Thus, the segment losses of energy in a partial admission stage are determined by a number of interrelated phenomena at the ends of nozzle segments. These losses are proportional to the quantity of displaced and sucked in stagnation steam, i.e. to the width and height of moving blades, they are also influeced by the velocity ratio, the number of end pairs of nozzle segments (non-closed), and the relative blade efficiency.

The sum of windage loss and segment loss constitutes the energy loss due to partial admission.

For governing stages with partitions between nozzle groups, the highest degree of partially (maximum) is 0.8 to 0.9 depending on the design of nozzle arcs. For intermediate stages with the optimal degree of partially (optimum) is greater than 1, it may often be taken that one to avoid high is taken that 1. Besides, with optimum greater than 0.7, it may often be taken that 1 to avoid high dynamic forces on moving blades which appear in partial admission stages.

(5) Disc Friction losses: If a disc similar to a rotor of the steam turbine rotates in a slightly viscous fluid like steam, there is a true surface friction loss due to relative motion between the disc and steam particles. Consequently the moving surface exerts a drag on the steam and this drag sets the steam in motion and produces a definite circulation of it whose nature depends on the shape of the space enclosing the disc.

The disc friction is a function of the diameter of the disc, peripheral speed of the disc and the density of the fluid. An optimum value of diameter of the disc, its speed and rotor casing clearance minimise the disc friction.

(6) Clearance loss: The nozzles are fixed in a stationary diaphgram and the moving blades are fixed in rotor. Since the stationary and rotating parts cannot be combined hence a clearance is left between the diaphragm and the shaft or hub. There exists a difference in pressure between the two sides of the diaphragm carrying the nozzles due to expansion of steam in nozzles. This difference in pressure causes the flow of steam through the clearance space, known as leakage of steam as this steam is not used for conversion into mechanical work. To reduce this leakage, special labyrith packings are employed. Labyrith packings are discussed in the next section.

In the reaction turbines the guides (fixed) blades are attached to the casing and the moving blades are attached to the drum or rotor. There is a gap between the guide blades and the rotor and the moving blades and the stator of the turbine. The pressure drop the leakage of steam through a radial gap and it produces loss of heat energy.

(7) Losses due to wetness of steam: Near the exit of the turbine usually operate under conditions at which steam is wet resulting in the formation of heterogeneous mixture of water droplet and steam, and both have different velocities. Water droplets may appear as a film running along the casing, as a fine mist suspended in the steam and a film on the nozzle blades. On the rotor blades droplets are flung off due to centrifugal force. In addition to this the steam particles given an accelerating force to the droplets of water resulting in reduction of kinetic energy of flowing steam.

Since there is a much difference in the specific volume of steam and water particules hence the acceleration i.e. The absolute velocity of steam is much more than that of water particles. This causes the deflection of water droplets on to the back of moving blades as a result of which the moving blades experience an impact force caused by the impingement of water droplets on their backs. These impacts on the back side of the moving blades produces erosion of blade. Moreover, some of the kinetic energy of the steam passing through the moving blades passages has to be spent in overcoming the impact of the water droplets which actually acts as a braking force.

In the region where the steam is wet, all the quantity of steam passing through the stage does not contribute the mechanical work only the dry portion of steam during the process of expansion does work. Thus the fraction of steam which condenses into water droplets during the process of expansion basically governs the losses resulting from the wetness of steam.

(8) Carry over loss : Carry over loss is due to the formation of eddies in the annular space between nozzles or fixed blades and moving blades. Due to this loss the kinetic energy of steam available at the succeeding row of moving blades for utilisation is less than the kinetic energy at the exit of the proceding row of nozzles or fixed blades.

(9) Exhaust loss : Exhaust losses are caused due to the steam leaves the turbine with finite absolute velocity which is partly or wholly lost. In the single stage turbine the losses are high about 10 to 12%. In multistage turbines this is much less and is equal to the absolute velocity of exhaust from the last stage.

(10) Radiation and conduction losses : The energy losses due to radiation are comparatively small in most cases and may generally be neglected.

2.5.2 External losses

External losses are the losses which do not affect the steam condition while it flows through the turbine. These losses include mechanical losses and losses due to the end leakage.

(1) Mechanical losses : These losses are caused due to frictional losses due to the bearings of the turbine shaft and the accessories coupled to the turbine. These losses reduces the power output of the turbine and are termed as mechanical losses.

(2) Leakage loss : Leakage occurs at the external glands where the turbine shaft passes through the casing. At the high pressure end the steam leaks into the atmosphere, and at the low pressure end i.e. at condenser end the air from the atmosphere leaks into the turbine.

2.6 SEALS

Seals are very important and critical component in turbomachinery, especially on high pressure and high speed equipment. We are discussing here the sealing system that covers principally sealing systems used between the rotor and stator elements of turbomachinery. They fall into two main categories: noncontacting seals and face seals.

Since these seals are an integral part of the rotor system, they affect the dynamic operating characteristics of the machine; for instance, both the stiffness and damping factors will be changed by seal geometry and pressure.

2.6.1 Labyrinth glands

Leakage losses of the fluid in turbine stages since part of the fluid moves through clearance and does not enter the nozzle or moving blade passages. This part of the working fluid performs no work, and therefore, the energy of the leaking steam is lost for the stage.

Steam leakages can occur owing to improper design of various joints of turbine elements, for instance, between halves of diaphragms, between mounting surfaces of diaphgrams, and between diaphragm yokes and their seals in the turbine casing.

Steam leakages through gaps between the stator and rotor of a turbine are minimized by mounting labyrinth glands which are essentially a sequence of constrictions for the flow of leaking steam.

The labyrinth is one of the simplest of sealing devices. It consists of a series of circumferential strips of metal extending from the shaft or from the bore of the shaft housing to form a cascade of annular orifices. Labyrinth seal leakage is greater than that of clearance bushings, contact seals, or film riding seals. Consequently, labyrinth seals are utilized when a small loss in efficiency can be tolerated. They are sometimes a valuable adjunct to the primary seal.

Advantages

The major advantages of labyrinth seals are their

1. Simplicity,
2. Reliability,
3. Tolerance to dirt,
4. System adaptability,
5. Very low shaft power consumption,
6. Material selection flexibility,
7. Minimal effect on rotor dynamics,
8. Back diffusion reduction,
9. Integration of pressure,
10. Lack of pressure limitations, and
11. Tolerance to gross thermal vibrations.

Disadvantages

The major disadvantages are the (1) High leakage, (2) Loss of machine efficiency, (3) Increased buffering costs, (4) Tolerance to ingestation of particulates with resulting damage to other critical items such as bearings, (5) The possibility of the cavity clogging due to low gas velocities or back diffusion and (6) The inability to provide a simple seal system that meets the standards.

Labyrinths seals are simple to manufacture and can be made from conventional materials. Early design of labyrinth seals used knife-edge seals and relatively large chambers or pockets between the knives. These relatively long knives are easily subject to damage. The modern, more functional, and more reliable labyrinth seals consist of sturdy, closely spaced lands (Fig. 2.10).

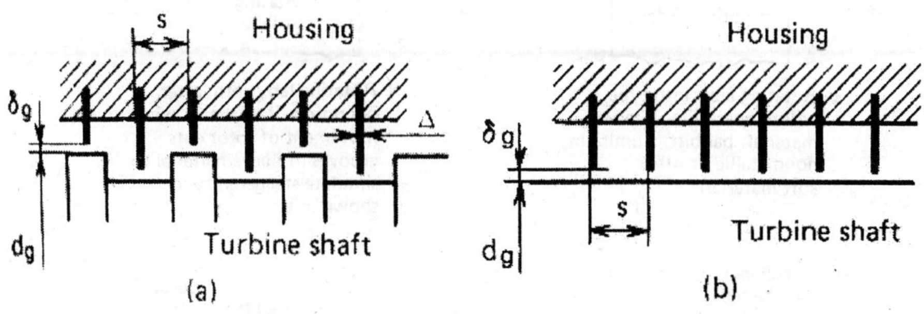

Fig. 2.10. Schemes of labyrinth glands of (a) stepped and (b) straight through type

Some labyrinth seals are shown in Fig. 2.11.

Fig. 2.11 (a) is the simplest form of the seal. Fig. 2.11 (b) shows a grooved seals is more difficult to manufacture but produces a tighter seal. Fig. 2.11 (c) and Fig. 2.11 (d) are rotating labyrinth type seals. Fig. 2.11 (e) shows a simple labyrinth seal with buffered gas for which pressure must be maintained above the process gas pressure and the outlet pressure (which can be greater than or less than the atmospheric pressure). The buffered gas produces a fluid barrier to the process gas. The educter sucks gas from the vent near the atmospheric end. Fig. 2.11 (f) shows a buffered, stepped labyrinth. The step labyrinth gives a tighter seals. The matching stationary seal is usually manufactured from soft materials such as babbitt or bronze, while the stationary or rotating labyrinth glands are made from steel. This composition enables the seal to be assembled with minimal clearance. The glands can therefore cut into the softer materials to provide the necessary running clearance for adjusting to the dynamic excursions of the rotor.

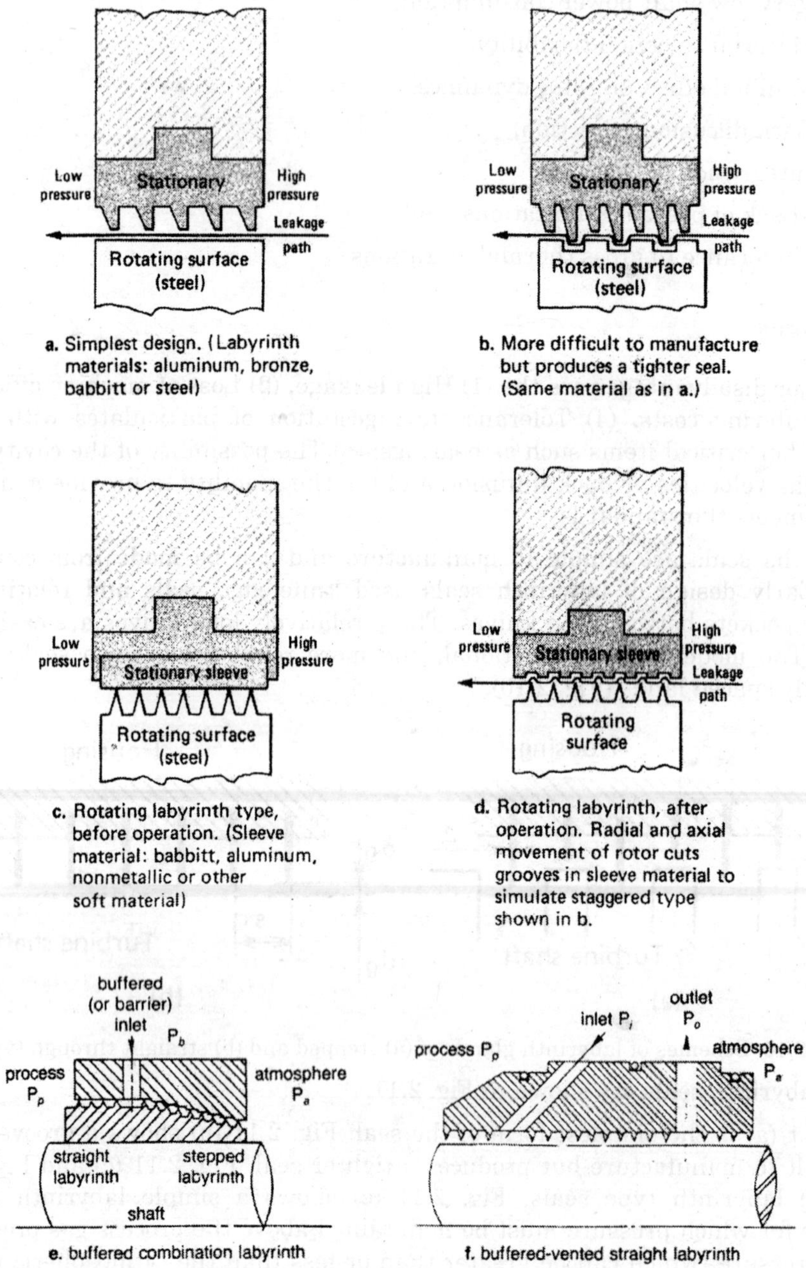

Fig. 2.11. Various configurations of labyrinth seals.

To maintain maximum sealing efficiency, it is essential that the labyrinth glands maintain sharp edges in the direction of the flow. This requirement is similar to that in orifice plates. A sharp edge provides for maximum vena contracta effect, and hence maximum restriction for the leakage flows (Fig. 2.12).

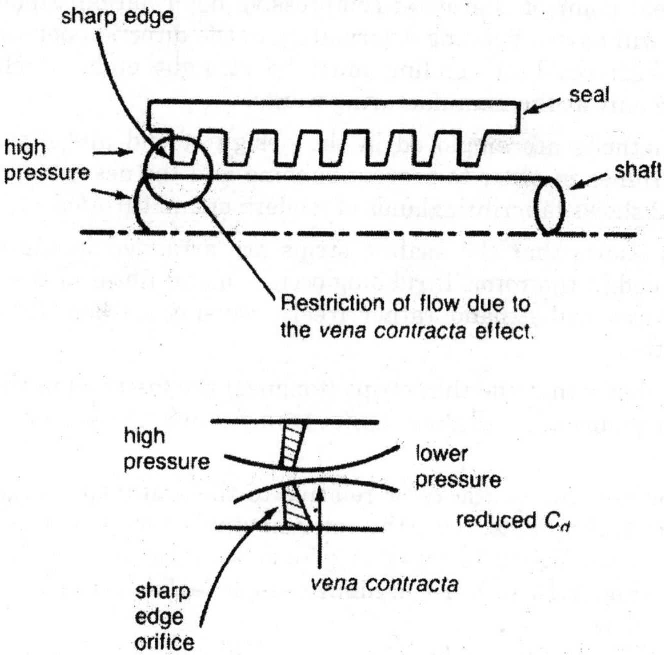

Fig. 2.12. Theory behind the knife edge arrangement

The leakage of a labyrinth seal can be kept to a minimum by providing :

(1) Minimum clearance between the seal glands and the seal sleeve.

(2) Sharp edges on the glands to reduce the flow discharge coefficient, and

(3) Grooves or steps in the flow path for reducing dynamic head carryover from stage to stage.

For effective functioning of any labyrinth gland, it is essential how its elementary step is designed, i.e. the slit formed by a sealing strip and the subsequent expansion chamber. It is important that the kinetic energy of steam jet ejected from the slit is extinguished fully in the expansion chamber. Under this condition, the steam leakage through the gland will be minimized.

It is desirable to increase the number of slits in a labyrinth gland in order to reduce the steam leakage through it. For that reason, desingers strive for accommodating as many slits (or strips) in a given length of gland as is feasible. Provided that the expansion chambers will have sufficient dimensions for full extinction of the kinetic energy of the flow after slits.

The flow rate of steam in a labyrinth gland can be reduced most effectively by leaving narrow clearance at sealing tongues between the stator and rotor. With very narrow clearances, however, rubbing between the rotor and stator may occur. If one sided rubbing takes place between the rotor and a sealing strip mounted in the stator

(Fig. 2.10), the heat liberated due to friction will heat up the surface layers of the rotor metal. Due to thermal expansion of the metal, the rotor buckles at the heated (rubbing) side, which in turn increases the rubbing effect and heating and buckling of the rotor. These deformations of the rotor cause compressive stresses in the rubbing region which can exceed by yield point of the metal compressive deformation. Upon shut down and cooling, the rotor will have a bending deformation in the direction opposite to the rubbing region. A rotor with residual bending must be straight ends which is an intricate operation possible only at the manufacturing works.

Two main methods are employed in the design of end and diaphragm glands of modern steam turbines in order to prevent rubbing and the associated residual bending of rotors. Fig. 2.12 shows labyrinth glands of modern steam turbines.

Fig. 2.12 (a) shows that the sealing strips are arranged in the stator, expansion grooves are provided in the rotor. If rubbing occurs, metal fibres of the rotor in a section between two grooves will expand rather freely, causing no bending of the rotor and residual deformation.

Fig. 2.12 (b) shows that the thin strips (tongues) are fastened in the rotor. Since the thickness of these strips is small they cannot transfer much heat from a rubbing zone to the rotor.

If axial displacements of the rotor relative to the stator are large, use is made of either straight through glands, with the rotor having no steps and with sealing strips arranged in the stator (Fig. 2.12 (c) or of glands with inclined sealing tonques on the rotor and stator (Fig. 2.12 (d)). To minimize steak leakage, the tonques are inclined against the steam flow.

Glands of the types shown in Fig. 2.12 (a) and (d), permit substantial (up to 35-35 mm) axial displacements of the rotor relative to the stator, whereas in stepped type glands (Fig. 2.12 (a) and (b)) axial displacements of the rotor are limited by the size of a groove in which the sealing strip can move.

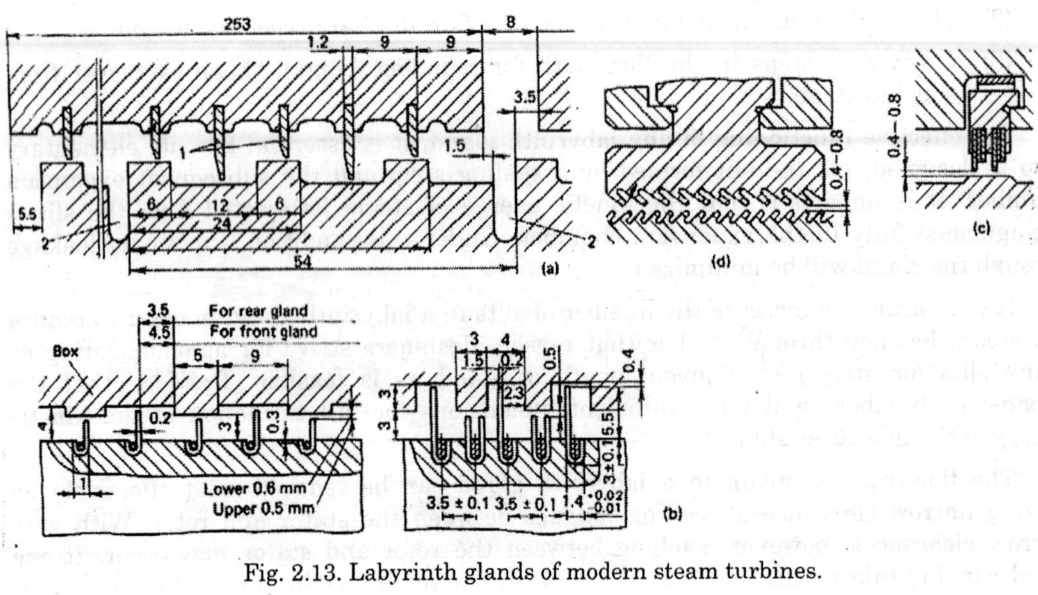

Fig. 2.13. Labyrinth glands of modern steam turbines.

2.7 CONSTRUCTIONAL DETAILS OF NOZZLES AND DIAPHRAGMS

The basic requirements of a nozzle are the following

1. Ease of manufacture and finishing and to obtain accurate channel sections at the HP end.
2. The wall surface should be smooth to reduce friction, particularly when steam velocity is high
3. Sudden changes in direction of the flowing steam especially at high velocities should be avoided.
4. The inlet of the nozzle should be designed to utilize the carryover energy from the previous stage to the greatest possible extent.
5. The conversion of thermal energy into kinetic energy should take place with greatest possible efficiency.
6. Nozzle and moving blades are designed according to the type of the turbine stage in which they will be arranged.

Fig. 2.14 shows the segment of nozzles suitable for the first stage of an impulse turbine.

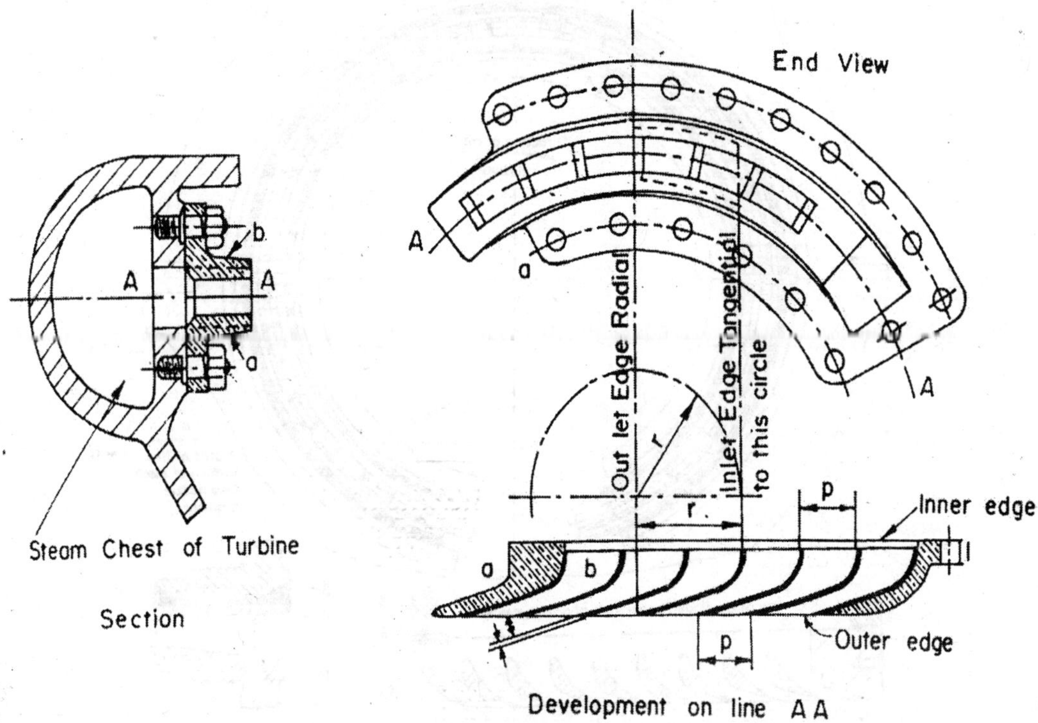

Fig. 2.14. A segment of nozzles suitable for the first stage of an impulse turbine

Fig. 2.15 shows the steam path of a two row high pressure governing stage of steam turbine and the nozzle segment of this stage. Surfaces of nozzle passages are nitrided to minimize erosive wear, especially of thin trailing edges of nozzle blades.

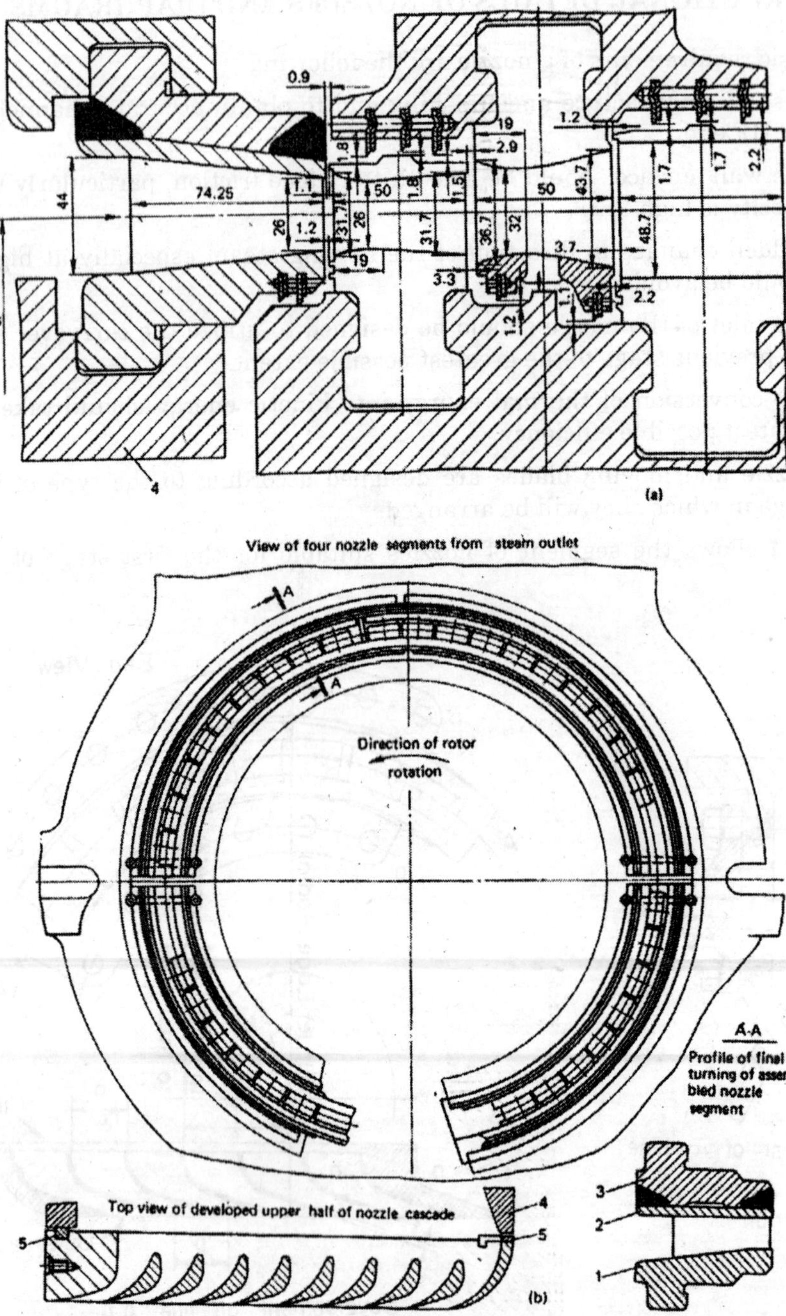

Fig. 2.15. Two row governing stage a - steam path; b - nozzle segment

The design of a nozzle segment of governing stage of steam turbine is illustrated in Fig. 2.16. The nozzle passages are divided by partitions into eight groups each of which is fed with steam from a separate governing valve. The extreme nozzle groups are closed by nozzle liners

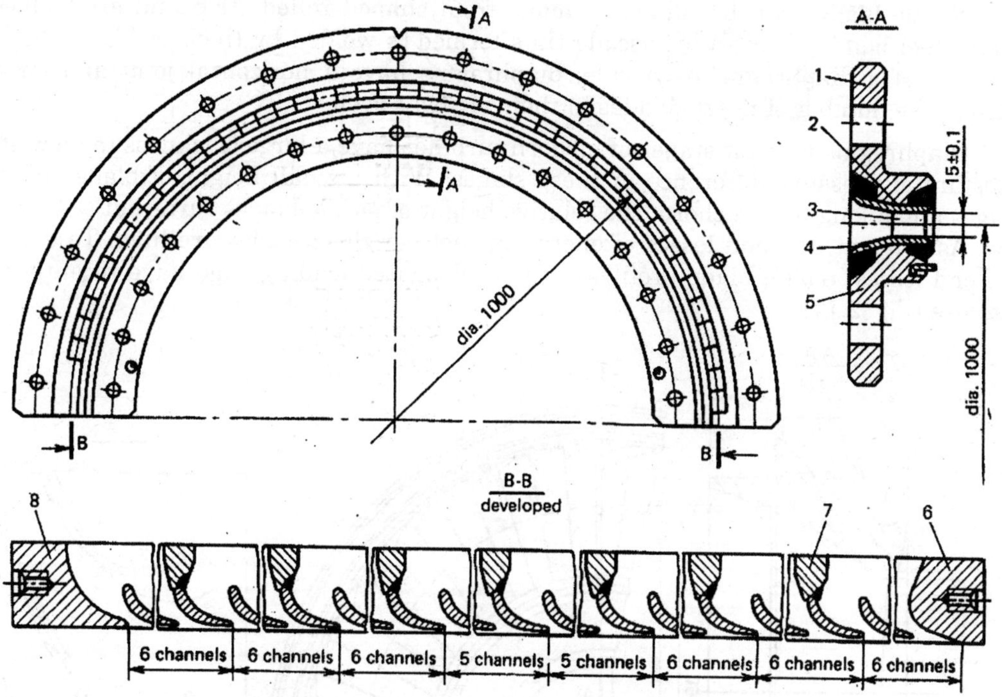

Fig. 2.16. Nozzle segment of governing stage.

Nozzle blades of non governing impulse type turbine stages are arranged in diaphragms. In modern steam turbines welded diaphragms are employed in turbine stages operating at steam temperatures above 200-250°C. An example of welded diaphragms is illustrated in Fig. 2.17.

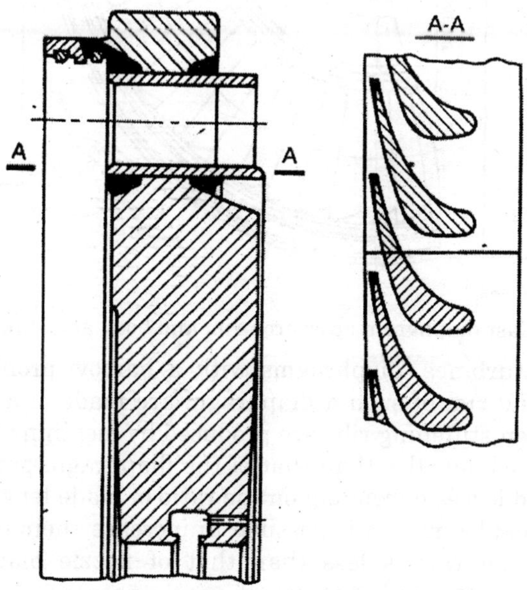

Fig. 2.17. Welded diaphragms.

Nozzle blades can be milled or made from shaped rolled strip and are welded to sheet-steel bands. The blade cascade thus formed is welded by thick welds to the body and rim of a diaphragm. As a rule, diaphragms have a horizontal joint and for that reason the number of nozzle blades in them is alwas even.

Diaphragms for first stages of modern turbines have a large thickness in view of the substantial pressure difference in these stages. With a small height of blades in these stages and a long profile chord, the relative height of nozzle blades turns out to be rather low. As a result, the end losses of energy in such nozzle cascades are high. To diminish the end losses, nozzle blade profiles with an elongated leading edge have recently come into use (Fig. 2.17).

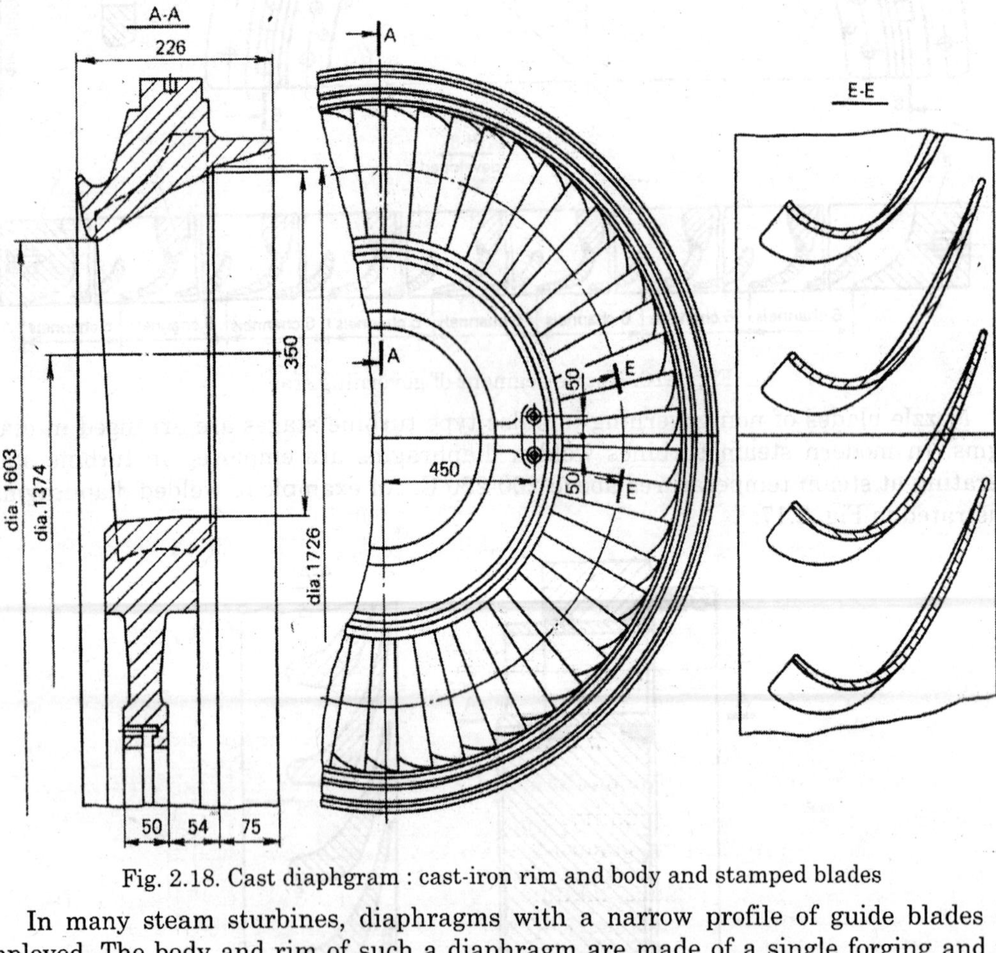

Fig. 2.18. Cast diaphgram : cast-iron rim and body and stamped blades

In many steam sturbines, diaphragms with a narrow profile of guide blades are employed. The body and rim of such a diaphragm are made of a single forging and the nozzle passages between stiffening ribs are produced by mchining (slottıng). The width of the nozzle blades is much smaller than that of the diaphragm proper, which is intended for diminishing the end losses depending on the ratio of blade length to profile chord. The use of narrow nozzle blades makes it possible to increase their relative height. Though the number of stiffening ribs is less than that of nozzle blades, a relatively thick boundary layer forms on the end surfaces of nozzle passages between the ribs, which decreases the total efficiency of the cascade. For that reason, the efficiency of both types of cascade (with wide nozzle blades and with narrow blades and wide diaphragm) turns out to be roughly the same and sometimes even lower for narrow blades. Narrow-blade diaphgrams are now employed only rarely.

Cast diaphragms find application in turbine stages operating in the region of low steam temperatures, for instance, in low-pressure parts of small-capacity turbines. In that case, nozzle blades are stamped from sheet steel and secured in the body and rim of a cast iron or steel diaphragm during casting of the latter (Fig. 2.18). For better fastening of blades in the casting, holes or slots are provided in their roots. Cast diaphragms of this design cannot ensure a high aerodynamic performance of nozzle passages because of the following reasons: (a) cast surfaces of meridional contours are rough; (b) casting technology fails to ensure high accuracy of blade adjustment and precise dimensions of blade passages; and (c) stamped nozzle blades cannot have an effective profile. For these reasons, milled and welded-stamped blades are preferably employed in the last stages of high capacity steam turbines of thermal and nuclear power stations (Fig. 2.19).

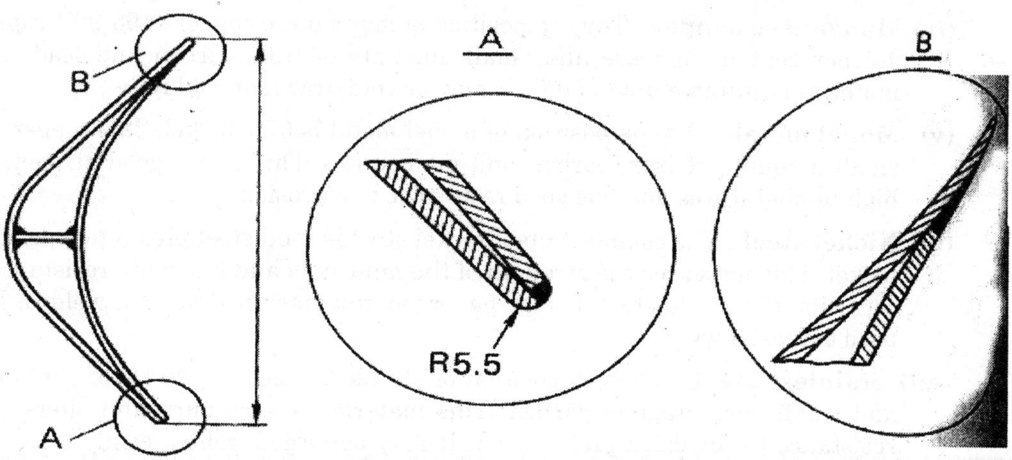

Fig. 2.19. Welded-stamped nozzle blade (cross-section)

2.8.1 Manufacturing methods

The most common methods of manufacturing turbine blades are given below.

(i) Cold drawing : This method is cheap and gives good finish. The material obtained is of uniform quality. Reaction blades, and plate and profile blades with separate spacers are generally manufactured in strip form by cold drawing.

(ii) Machining : Blades manufactured by machining are stronger than rolled blades, but are costlier. Impulse blades with integral spacers are machined from bars.

(iii) Drop forging : Long blades of complicated shapes such as tapered or twisted shapes are made from drop forging. The disadvantage of drop forging is that it does not give uniform strength.

(iv) Rolling: Rolling may be hot or cold. Impulse and reaction blades with separate spacers are manufactured from rolling.

2.8.2 Material

The selection of material, besides desirable properties, largely depends on cost. The following are the materials which are generally used for blades.

(i) **Brass :** The composition of brass used for blades is 70-72% copper and 28-30% zinc, Brass blades are cold drawn. These blades are suitable for temperatures upto 225°C. It is resistant to corrosion. These days most turbines work at higher temperatures so brass blades are not used except in small turbines.

(ii) **Nickle brass :** The composition of nickle brass is 50% copper, 40% zinc and 10% nickle. The temperature limit is 200-225°C. It may be cold drawn.

(iii) **Phosphor bronze :** The composition of phosphor bronze is copper 86-92%, tin 14-8% with a small percentage of phosphorous, say 1% Phosphorous acts as deoxidizer and prevents corrosion. Phosphor bronze may be forged, cold drawn and cold rolled. It is suitable below temperature 325°C.

(iv) **Manganese copper :** The composition of manganese copper is 95-96% copper, 4-5 per cent manganese and small amounts of iron, carbon and lead. This material is suitable upto 350°C. It may be cold drawn or cold rolled.

(v) **Monel metal :** The composition of monel metal is 67% nickel, 28% copper and small amounts of iron, carbon and manganese. This has a good strength at high temperatures and has good resistance to corrosion.

(vi) **Nickel steel :** The composition of nickel steel is mild steel plus 3 to 5 per cent nickel. This increases the strength of the mild steel and has more resistance to corrosion than mild steel. It may be forged and machined but not welded. It is used extensively.

(vii) **Stainless steel :** The composition of stainless steel is 12-14% of chromium and small percentage of carbon. This material is very hard and offers good resistance to corrosion and erosion. It may be forged, rolled, easily machined and welded.

2.8.3 Mounting turbine moving blades

Moving blades of turbine stages are mounted on the turbine rotor. In impulse type stages, each moving-blade cascade is composed of moving blades set into the rim of a disc on the rotor. The disc rim has a milled slot in which the roots of moving blades are fastened. The required spacing between the blades is obtained by means of root spacers which, as a rule, are milled integral with blade roots. At the periphery of a moving blade cascade, blade passages are usually covered by a band which is either riveted to the tips of blades or milled integral with blades. The design embodiment of moving blades depends substantially on the forces acting on them. Firstly, there are centrifugal forces acting on the blade body. These forces increase with an increase of the average diameter of the stage and of the length of moving blades. In short blades of the initial stages, centrifugal forces are relatively low, but attain high values in the last stages of high capacity condensing turbines.

Of large importance are also the forces of the steam flow which create a bending moment on the blades. The highest bending stresses appear, as a rule, in root sections of blades. Therefore these sections are the critical ones, since the bending stresses are combined here with the stresses due to centrifugal forces, which are also the highest in these sections. Owing to rotation of moving blades, the appearance of edge traces, and partial admission of steam in governing stages, the moving blades are subjected to a periodically varying bending moment which creates the danger of vibrational breakage of blades.

Example of design of short moving blades in which centrifugal stresses are relatively low are shown in Figs. 2.20, 2.21 and 2.22. The blades shown in Fig. 2.20 are of the reaction type. The blade is made from rolled shaped strip and has small forged shoulders at both sides of its root. It is fastened in the rotor slot by these shoulders by means of separate root spacers with serrations.

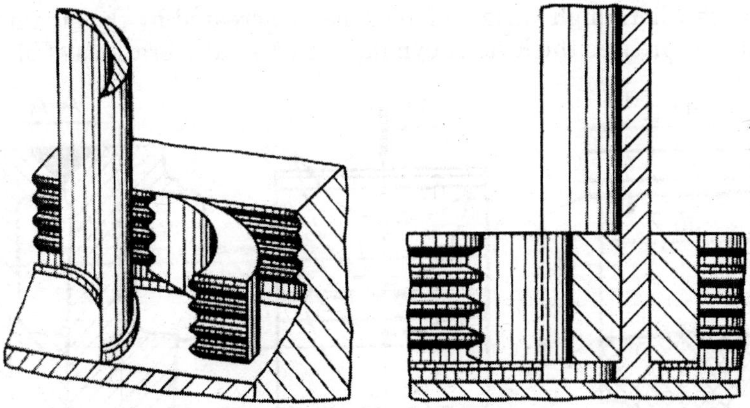

Fig. 2.20. Reaction blades made from rolled profiled strip.

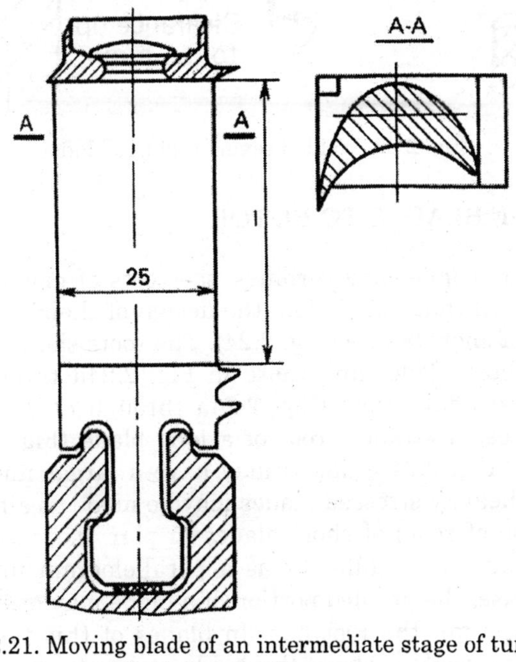

Fig. 2.21. Moving blade of an intermediate stage of turbine, milled integral with root spacer

Figure 2.21 illustrates a milled blade whose root is made integral with root spacer. The root has a rectangular form when seen from the blade top. Milled blades are characterized by an elevated strength. As seen in the figure, the band which is riveted to the blades has two radial and one axial tongue for narrowing the clearances between the band and stator elements. The blade root is also seen to have two tongues to seal the root clearance between the diaphgram and blades of the stage. The band forms the outside boundary of the flow in blade passage. A metal pad is caulked in under the blade root to fix the blade in proper position in the disc slot. In the upper portion of the blade root there are two stoppers to minimize bending stresses in the jaws of the disc rim. Fig. 2.22 shows a blade milled integral with the root and band. In order to increase the bending rigidity, such blades are welded together pairwise in the band sections. Blades of this type are employed in turbine stages with partial steam admission, for instance, in governing stages where high variable forces may be created by the steam flow. The high bending rigidity decreases the level of dynamic stresses at vibrations of blades.

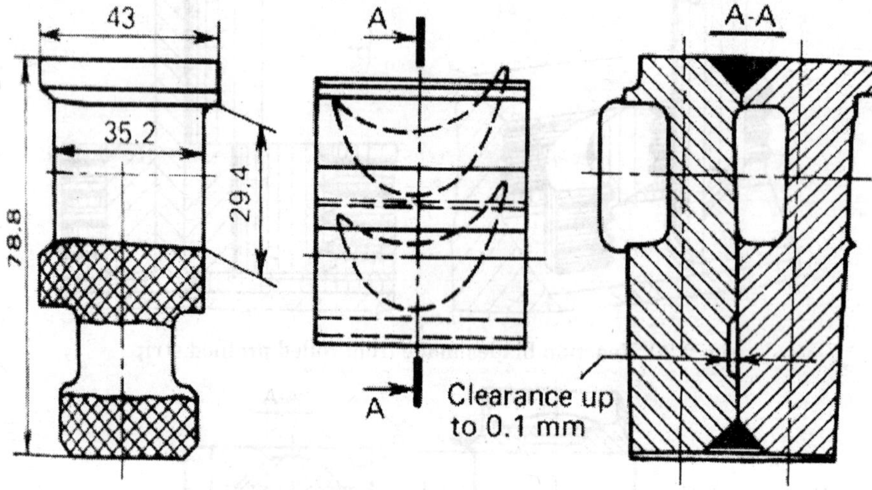

Fig. 2.22. Welded segment of two blades

2.9 ATTACHMENT OF BLADES TO ROTOR

In the last stages of condensing turbines, the roots of moving blades are subject to high centrifugal forces. In that connection, the design of their roots differs substantially from that of the roots of short blades (Fig. 2.24). The roots shown in Fig. 2.24a through c and g are typical of short blades and those in Fig. 2.24d through f, h, and i, of long blades. T-shaped and straddle roots (Fig. 2.24a through c) have a bearing surface to absorb centrifugal forces. A straddle root of a long blade (Fig. 2.24 d) has three such surfaces. Forked roots (Fig. 2.24 e and f) may have a larger number of prongs in long blades or the prong in heavily stressed blades may be made as a body of equal resistance (Fig. 2.24 f). The shape of roots of short blades in plan (i.e. viewed in the direction of blade axis) is often made rectangular or as a parallelogram in order to simplify root manufacture. In that case, the profiled portion of a blade may overhang the root, i.e. may have larger dimensions than the root has. In blades of this type, local stress concentration can occur in the section where the blade connects the root. In heavily loaded moving blades, this stress concentration is inadmissible and for that reason their roots have a shape which avoids overhang of blade edges relative to the root which is then of the stepped type (Fig. 2.24 f) or herring-bone type with circular insertion (Fig. 2.24 i).

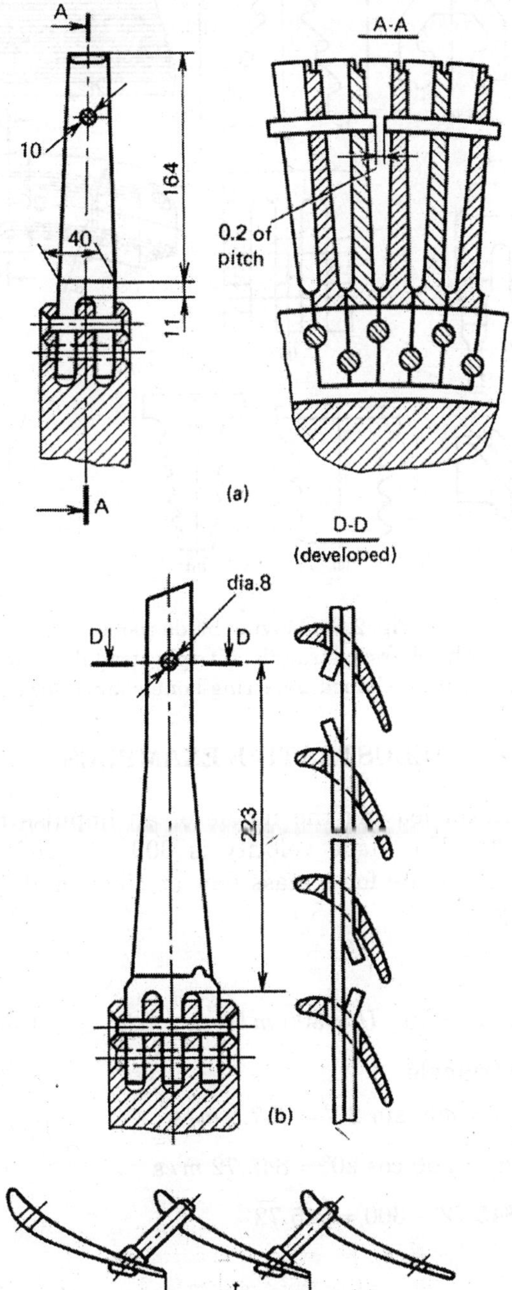

Fig. 2.23. Large-length moving blades with wire ties :
a - rigid; b - damping; c-z- shaped.

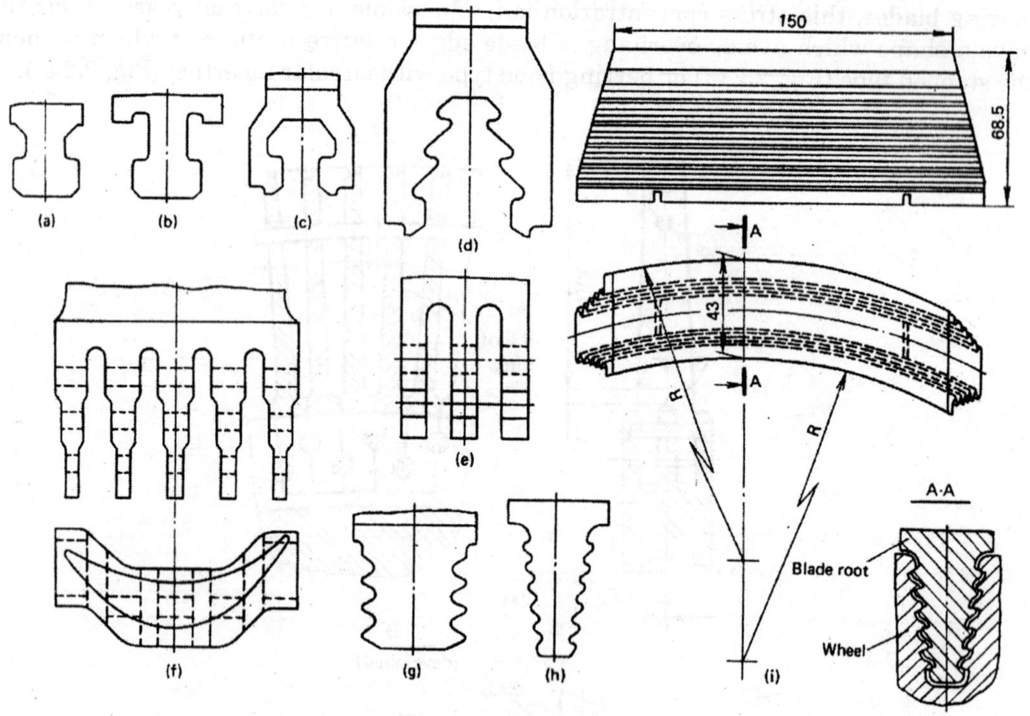

Fig. 2.24. Moving blade roots :
a - T-shaped; b- T-shaped with collars; c-straddle; d-triple straddle; e-forked; f- forked stepped; g - serrated; h- herring-bone-shaped; i-herring-bone-shaped for circular insertion.

ILLUSTRATIVE EXAMPLES

2.1. The velocity of steam leaving the nozzle of an impulse turbine is 900 m/s and nozzle angle is 20°. The blade velocity is 300 m/s and and the blade velocity coefficient is 0.7. Calculate for a mass flow of 1 kg/s and symmetrical blading the diagram efficiency. **[JNVU, 1990]**

Solution : Given :

$$C_1 = 900\ m/s,\ \ \alpha_1 = 20°, U = 300\ m/s\ \ k = 0.7,\ \ \dot{m} = 1\ kg/s$$

From inlet velocity triangle

$$C_{a1} = C_1 \sin \alpha_1 = 900 \sin 20° = 307.8\ m/s$$

$$C_{w1} = C_1 \cos \alpha_1 = 900 \cos 20° = 845.72\ m/s$$

$$\therefore C_{w1} - U = 845.72 - 300 = 545.72$$

Now

$$\tan \beta_1 = \frac{C_{a1}}{C_{w1} - U} = \frac{307.8}{545.72} = 0.5640255$$

$$\therefore \beta_1 = \tan^{-1}(0.5640255) = 29°25'$$

$$V_1 = \frac{C_{a1}}{\sin\beta_1} = \frac{307.8}{\sin 29°25'} = \frac{307.8}{0.4912702} = 626.54\ m/s$$

$$V_2 = K\,V_1 = 0.7 \times 626.54 = 438.58\ m/s$$

Given $\beta_1 = \beta_2 = 29°25' = \beta$

$$C_{w2} = V_2 \cos\beta - U = 438.58 \cos 29°25' - 300$$

$$= 382 - 300 = 82\ m/s$$

$$\therefore \Delta C_w = C_{w1} - (-C_{w2}) = 845.72 + 82$$

$$= 927.72\ m/s$$

$$= \frac{\dot{m}\,\Delta C_w \times U}{1000}$$

$$= \frac{1 \times 927.72 \times 300}{1000} = 278.316\ kW$$ Ans.

(e) The diagram efficiency

$$\eta_d = \frac{\dot{m}\,\Delta C_w\,U}{C_1^2/2} = \frac{278316 \times 2}{(900)^2}$$

$$= 0.6872 = 68.72\%$$ Ans.

2.2 The mean diameter of the blades of an impulse turbine with a single row wheel is 1 m and the speed is 3000 rpm. The nozzle angle is 18°, the ratio of linear velocity of blade to absolute velocity of steam entering moving blade is 0.42, and the relative velocity at outlet from the blades to that at inlet is 0.84. The outlet angle of the blade is to be made 3° less than the inlet angle of the blade. The steam rate of flow is 8 kg/s. Determine the blading efficiency. **[JNVU, 1998]**

Solution : Given :

$$D = 1m,\quad N = 3000\ rpm\quad \alpha_1 = 18°,\quad \rho = \frac{U}{C_1} = 0.42$$

$$\frac{V_2}{V_1} = 0.84,\quad \beta_2 = \beta_1 - 3°\quad \dot{m} = 8\ kg/s$$

$$U = \frac{\pi DN}{60} = \frac{\pi \times 1 \times 3000}{60} = 157.1\ m/s$$

$$\frac{U}{C_1} = 0.42 \qquad \therefore C_1 = \frac{157.1}{0.42} = 374\ m/s$$

$$C_{w1} = C_1 \cos\alpha_1 = 374 \cos 18^\circ = 355.7\ m/s$$

$$C_{a1} = C_1 \sin\alpha_1 = 374 \sin 18^\circ = 115.57\ m/s$$

$$\tan\beta_1 = \frac{C_{a1}}{C_{w1} - U} = \frac{115.57}{355.7 - 157.1} = \frac{115.57}{198.6} = 0.58$$

$$\therefore \beta_1 = \tan^{-1}(0.58) = 30.196^\circ$$

$$\therefore \beta_2 = \beta_1 - 3^\circ = 30.196 - 3 = 27.196^\circ$$

$$V_1 \cos\beta_1 = C_{w1} - U = 198.6$$

$$\therefore V_1 = \frac{198.6}{\cos 30.196^\circ} = 229.78\ m/s$$

$$V_2 = 0.84 \times 229.78 = 193\ m/s$$

$$C_{a2} = V_2 \sin\beta_2 = 193 \sin 27.196^\circ = 88.2\ m/s$$

$$V_2 \cos\beta_2 = U + C_{w2}$$

$$193 \cos 27.196^\circ = 157.1 + C_{w2}$$

$$\therefore C_{w2} = 171.676 - 157.1 = 14.576\ m/s$$

$$\Delta C_w = C_{w1} + C_{w2} = 355.7 + 14.576 = 370.276$$

$$\Delta C_a = C_{a1} - C_{a2} = 115.57 - 88.2 = 27.37$$

$$\eta = \frac{2\,\Delta C_w\, U}{C_1^2} = \frac{2 \times 370.276 \times 157.1}{(374)^2}$$

$$= 0.83174 = 83.174\%$$ **Ans.**

2.3 A single row, impulse turbine has blades whose inlet angle is 40° and exit angle 37°. The mean blade speed is 230 m/s and the nozzles are inclined at an angle of 27° to the plane of rotation of the blades. There is a 10% loss of relative velocity due to friction in the blades. The turbine uses 550 kg/hr of steam. Determine the diagram efficiency. **[JNVU, 2005]**

Solution : Given :

$\alpha_1 = 27^\circ$, $\beta_1 = 40^\circ$, $\beta_2 = 37^\circ$, $U = 230\ m/s$

From inlet velocity triangle

Using sine rule

$$\frac{U}{\sin 13^\circ} = \frac{C_1}{\sin 140^\circ} = \frac{V_1}{\sin 27^\circ}$$

$$\therefore C_1 = \frac{U \sin 140^\circ}{\sin 13^\circ} = \frac{230 \times 0.64278}{0.22495} = 657.21\ m/s$$ **Ans.**

$$V_1 = \frac{U \sin 27^\circ}{\sin 13^\circ} = \frac{230 \times 0.4539}{0.22495} = 464.18\ m/s$$

$$\therefore\ V_2 = 0.9\ V_1 = 417.762\ \ m/s$$

$$C_{a1} = V_1 \sin\ \beta_1 = 464.18\ \sin 40^\circ$$

$$C_{w1} = C_1 \cos 27^\circ = 657.21\ \cos 20^\circ = 585.578\ m/s$$

From outlet velocity triangle

$$C_{w2} = V_2 \cos\ \beta_2 - U$$

$$= 417.762 \cos 37^\circ - 230$$

$$= 333.639 - 230$$

$$= 103.639\ \ m/s$$

$$\Delta\ C_w = C_{w1} + C_{w2} = 585.578 + 103.639 = 689.217$$

$$\eta_d = \frac{2\ \Delta\ C_w\ U}{C_1^2} = \frac{2 \times 689.217 \times 230}{(657.21)^2} = 0.734 = 73.4\%$$ Ans.

2.4 In a single stage impulse turbine the nozzle discharge the fluid on the the blades at an angle of 25° to the plane of rotation and the fluid leaves the blades with an absolute velocity of 300 m/s at an angle of 120° to the direction of motion of the blades. If the blades have equal inlet and outlet angles and there is no axial thrust, estimate find the diagram efficiency. **[JNVU 1996, 1997, 1999, 2000]**

Solution : Given,

$$\alpha_1 = 25^\circ \qquad C_2 = 300\ m/s, \qquad \alpha_2 = 180 - 120 = 60^\circ$$

$$C_{a1} = C_{a2}\ (\text{no axial thrust})$$

From outlet velocity triangle

$$C_{a2} = C_2 \sin\ \alpha_2 = 300 \sin 60^\circ = 259.8\ m/s$$

$$C_{w2} = C_2 \cos\ \alpha_2 = 300 \cos 60^\circ = 150\ m/s$$

From the given condition

$$C_{a1} = C_{a2} = 259.8 = C_1 \sin\ \alpha_1$$

$$\therefore\ C_1 = \frac{259.8}{\sin 25^\circ} = 614.74\ m/s$$

$$C_{w1} = C_1 \cos\ \alpha_1 = 614.74\ \cos 25^\circ = 557.142\ m/s$$

$$\Delta\ C_w = C_{w1} + C_{w2} = 557.142 + 150 = 707.142\ m/s$$

Diagram efficiency

$$\eta_d = \frac{2\ \Delta\ C_w\ U}{C_1^2}$$

$$= \frac{2 \times 707.142 \times 203.571}{(614.74)^2} = 0.7618 = 76.18\% \qquad \text{Ans.}$$

2.5 A single row impulse turbine received 3 kg/s steam with velocity of 425 m/s. The ratio of blade speed to jet speed is 0.4 and the stage output is 169 kW. If the internal losses due to disc friction etc. amount to 14.7 kW, determine the blading efficiency, the velocity coefficiency and the leaving loss. The nozzle angle is 19° and the blade exit angle is 17°. **[JNVU 2006]**

Solution : Given :

$$\dot{m} = 3\ kg/s, \qquad C_1 = 425\ m/s, \qquad \rho = \frac{U}{C_1} = 0.4 \text{ Stage output } = 169\ kW$$

Internal losses due to disc friction = 14.7 kW

$$\therefore \text{Total power} = 168 + 14.7 = 183.7\ kW$$

$$\alpha_1 = 19^\circ \text{ and } \beta_2 = 17^\circ$$

From the inlet velocity traingle

$$C_{w1} = C_1 \cos\alpha_1 = 425 \cos 19^\circ = 401.85\ m/s$$

$$C_{a1} = C_1 \sin\alpha_1 = 425 \sin 19^\circ = 138.4\ m/s$$

$$\frac{U}{C_1} = 0.4 \ \therefore U = 425 \times 0.4 = 170\ m/s$$

$$C_{w1} - U = 401.85 - 170 = 231.85\ m/s$$

$$\tan\beta_1 = \frac{C_{a1}}{C_{w1} - U} = \frac{138.4}{231.85} = 0.5969$$

$$\therefore \beta_1 = 30^\circ 50'$$

Now $V_1 \cos\beta_1 = C_{w1} - U = 231.85$

$$\therefore V_1 = \frac{231.85}{\cos 30^\circ 50'} = 270\ m/s$$

$$P = \frac{\dot{m}\,\Delta C_w\,U}{1000}$$

$$183.7 = \frac{3 \times \Delta C_w \times 170}{1000}$$

$$\therefore \Delta C_w = 360.2$$

$$\therefore C_{w2} = \Delta C_w - C_{w1} = 360.2 - 401.85$$

$$= -41.65$$

From outlet velocity triangle

$$V_2 \cos\beta_2 = U - C_{w2}$$

$$= 170 - 41.65 = 128.35$$

$$\therefore V_2 = \frac{128.35}{\cos 17^\circ} = 134.2$$

$$\therefore \frac{V_2}{V_1} = \frac{134.2}{270} = 0.497 = 0.5 \qquad \text{Ans.}$$

$$C_{a2} = V_2 \sin \beta_2$$

$$= 134.2 \sin 17^\circ = 39.24\ m/s$$

Therefore,

$$C_2^2 = (C_{a2})^2 + (C_{w2})^2$$

$$= (39.24)^2 + (-41.65)^2$$

$$= 1539.82 + 1734.72 = 3274.54$$

$$\therefore C_2 = 57.22 \text{ m/s}$$

$$\therefore \quad \text{Leaving loss} = \frac{\dot{m}\, C_2^2}{2 \times 1000}$$

$$= \frac{3 \times 3274.54}{2 \times 1000} = 4.91\ kJ/s \qquad \text{Ans.}$$

$$\eta_d = \frac{2\,\Delta\, C_w\, U}{C_1^2} = \frac{2 \times 360.2 \times 170}{(425)^2} = 0.6788 = 67.88\% \qquad \text{Ans.}$$

2.6 A single row impulse turbine stage receives 5 kg of dry saturated steam per second at 10 bar. The steam is expanded in the nozzle to a pressure of 7 bar, with an efficiency of 0.95, and discharges at an angle of 20° to the plane of rotation of the blades. If the steam is to discharge from the blades in an axial direction, allowing a blade velocity coefficient of 0.88 and a blade speed of 150 m/s, find a suitable outlet blade angle. If the internal losses due to disc friction and windage amounts to 5 kJ/kg of steam, find the power developed. Also calculate (i) Leaving loss, (ii) Blade friction loss and (iii) Energy loss in nozzle in terms of kW. **[JNVU, 2009]**

Solution : Given

$$\dot{m} = 5\ kg/s, \qquad U = 150\ m/s, \quad \alpha_1 = 20^\circ, \quad C_{w2} = 0$$

$$V_1^2 = 0.88, \quad k = 0.95$$

From Mollier chart, isentropic enthalpy drop, $\Delta h = 67.5\ kJ/\text{kg}$

$$C_1 = 44.72\sqrt{K\ \Delta h}$$

$$= 44.72\sqrt{0.95 \times 67.5} = 358.11\ m/s$$

From inlet velocity triangle

$$C_{w1} = C_1 \cos \alpha_1 = 358.11 \cos 20^\circ = 336.5125 \; m/s$$

$$C_{a1} = C_1 \sin \alpha_1 = 358.11 \sin 20^\circ = 122.48 \; m/s$$

$$\tan \beta_1 = \frac{C_{a1}}{C_{w1} - U} = \frac{122.48}{336.5125 - 150} = \frac{122.48}{186.5125} = 0.6566$$

$$\therefore \beta_1 = \tan^{-1}(0.6566) = 33.3^\circ$$

$$C_{w1} = V_1 \cos \beta_1 + U$$

$$V_1 = \frac{C_{w1} - U}{\cos \beta_1} = \frac{336.5125 - 150}{\cos 33.3^\circ}$$

$$= \frac{186.5125}{0.8358} = 223.13 \; m/s$$

$$V_2 = 0.88 \quad V_1 = 0.88 \times 223.13 = 196.35$$

From outlet velocity triangle (Given $C_{w2} = 0$)

$$V_2 \cos \beta_2 = U$$

$$\cos \beta_2 = \frac{150}{196.35} = 0.7639$$

$$\therefore \beta_2 = \cos^{-1}(0.7639) = 40.187^\circ$$ Ans.

$$\tan \beta_2 = \frac{C_2}{U}$$

$$\tan 40.187 = \frac{C_2}{150}$$

$$\therefore = 150 \times 0.8446 = 126.7 \; m/s$$

$$\Delta C_w = C_{w1} - C_{w2} = 336.5125 - 0 = 336.5125 \; m/s$$

$$P = \frac{\dot{m} \, \Delta C_w \, U}{1000} = \frac{5 \times 336.5125 \times 150}{1000} = 252.38 \; kW$$

$$= 252.38 \; kW$$ Ans.

$$\text{Leaving loss} = \frac{C_2^2}{2000} = \frac{(126.7)^2}{2000} = 8.026 \; kJ/kg$$

$$= 5 \times 8.026 = 40.132 \; kW$$ Ans.

Energy loss in blade friction

$$= \frac{V_1^2 \left(1 - K^2\right)}{2000} = \frac{223.15^2 \left(1 - (0.882)^2\right)}{2000} = 49795.92 \times \frac{(1 - 0.7744)}{2000}$$

$$= 5.616 \; kJ/kg$$

$= 5 \times 5.616 = 28.08 \ kW$ Ans.

Energy loss in nozzle $= 0.05 \times 67.5$

$= 3.375 \ kJ/kg$

$= 5 \times 3.375$

$= 16.875 \ kW$ Ans.

2.7 A reaction steam turbine have blades with effective inlet and outlet angles of 80° and 20° respectively. The mean diameter of the blade rows is 1.2 m and the speed of rotation is 3000 rpm. Steam is admitted to the stage at 12 bar pressure and 200°C temperature and undergoes an adiabatic heat drop 48 kJ/kg. Ten percent of the steam is lost through tip leakage. If the pressure developed in the stage is 625 kW, determine (a) the stage efficiency, (b) the blade height.

[JNVU 1991, 1993, 1998, 2003, 2007, 2009]

Solution : Given

$$\alpha_1 = 20^\circ = \beta_2, \ \beta_1 = 80^\circ = \alpha_2 \ P = 625 \ kW, \ N = 3000 \ rpm, \ D = 1.2 \ m$$

$$\Delta h = 48 \ kJ/kg, \ tip \ leakage = 10\%$$

$$U = \frac{\pi D N}{60} = \frac{\pi \times 1.2 \times 3000}{60} = 188.5 \ m/s$$

Using sine rule in the inlet velocity triangle

$$\frac{C_1}{\sin(180-80)} = \frac{U}{\sin 60^\circ} = \frac{V_1}{\sin 20^\circ}$$

$$\therefore C_1 = 217.65 \sin 100^\circ = 214.35 \ m/s$$

$$V_1 = 217.65 \sin 20^\circ = 74.44 \ m/s$$

$$C_{a1} = C_1 \sin 20^\circ$$

$$= 214.35 \sin 20^\circ = 73.31 \ m/s$$

$$C_{w1} = C_1 \cos \alpha_1$$

$$= 214.35 \cos 20^\circ = 201.42 \ m/s$$

$$\Delta C_w = 2 C_1 \cos \alpha_1 - U$$

$$= 2 \times 201.42 - 188.5$$

$$= 402.846 - 188.5 = 214.35 \ m/s$$

$$P = \frac{\dot{m} \Delta C_w U}{1000}$$

$$625 = \frac{0.9 \dot{m} \times 214.35 \times 188.5}{1000}$$

$$\dot{m} = 17.19 \ kg/s$$

Using steam tables

At 1.2 bar, 200°C $v = 0.1693\ m^3/kg$

Therefore,

$$\dot{m} = \frac{\pi\, D\, h\, C_{a1}}{v_1}$$

$$17.19 = \frac{\pi \times 1.2 \times h \times 73.31}{0.1693}$$

$$h = 0.0105 \quad m \;= 10.5\ mm$$ Ans.

$$\eta_s = \frac{\Delta\, C_w\, U}{\Delta\, h}$$

$$= \frac{214.35\ \times 188.5}{48 \times 1000} = 0.8417 = 84.17\%$$ Ans.

2.8 A certain stage of Parson's turbine consists of one row of fixed blades and one row of moving blades. The details of the turbine are as follows. The mean diameter of the blades = 680 mm, RPM = 3000. The mass of steam passing per second = 13.5 kg, steam velocity at exit from fixed blades = 143.7 m/s. The blade outlet angle = 20°. Calculate the power developed in the stage and the gross efficiency assuming carryover coefficient as 0.74 and the efficiency of conversion of heat energy into kinetic energy in the blade channels as 0.92. **[JNVU 1993, 1998, 2001, 2010]**

Solution : Given

$$D = 0.68\ m,\ N = 3000\ rpm,\ \dot{m} = 13.5\ kg/s,$$

$$V_2 = C_1 = 143.7\ m/s,\ \alpha_1 = \beta_2 = 20^\circ,\ \phi = 0.74,\ \eta_c = 0.92$$

$$U = \frac{\pi\ D\ N}{60} = \frac{\pi \times 0.68 \times 3000}{60} = 106.81\ m/s$$

$$C_{w1} = C_1\ \cos 20^\circ = 143.7\ \cos 20^\circ = 135.03\ m/s$$

$$C_{a1} = C_1\ \sin 20^\circ = 143.7\ \sin 20^\circ = 49.148\ m/s$$

$$\therefore\ C_{w1} - U = 135.03 - 106.81 = 28.216$$

$$\tan \beta_1 = \frac{C_{a1}}{C_{w1} - U} = \frac{49.148}{28.216} = 17.418$$

$$\beta_1 = \tan^{-1}(1.7418) = 60.139^\circ$$

$$C_2 = V_1 = \frac{C_{a1}}{\sin \beta_1} = \frac{49.148}{\sin 60.139^\circ} = 56.672\ m/s$$

$$\Delta\ C_w = 2\ C_1\ \cos \alpha_1 - U = 2 \times 135.03 - 106.814 = 163.246$$

$$P = \frac{\dot{m}\ \Delta\ C_w\ U}{1000} = \frac{13.5\ \times 163.246 \times 106.814}{1000}$$

$= 235.34\ kW$ Ans.

$$\text{Enthalpy drop} = \frac{2\left(C_1^2 - \phi\ C_2^2\right)}{2\eta \times 1000} = \Delta\ h$$

$$= \frac{20649.69 - 2376.67}{0.92 \times 1000} = 19.86\ kJ/kg$$

$$\text{Gross efficiency} = \frac{\Delta\ C_w\ U}{\Delta\ h}$$

$$= \frac{163.246 \times 106.814}{19.86 \times 1000} = 0.8779 = 87.79\%$$ Ans.

2.9 A stage of a steam turbine with Parson's blading delivers dry saturated steam at 2.7 bar from the fixed blades at 90 m/s. The mean blade height is 40 mm, and the moving blade exit angle is 20°. The axial velocity of the steam is three quarters of the blade velocity at the mean radius. Steam is supplied to the stage at the rate of 2.5 kg/s. The effect of the blade tip thickness on the annular area can be neglected. Calculate:

(a) the rotational speed of the wheel,

(b) the diagram power,

(c) the diagram efficiency, and

(d) the enthalpy drop of the steam in this stage.

[JNVU, 1987, 2000, 2004, 2007, 2009]

Solution : Given

$C_{a1} = 0.75\,U,\ C_1 = 90\ m/s = V_2,\ \beta_2 = \alpha_1 = 20°,\ \dot{m} = 2.5\ kg/s,\ \ h = 0.04$

$C_{a1} = C_1\ \sin\ \alpha_1 = 90\ \sin\ 20° = 30.78\ m/s$

$$\therefore U = \frac{30.78}{0.75} = 41.04\ m/s$$

Using steam stables, at 2.7 bar

$$v_g = 0.6686\ m^3/kg = v$$

$$\dot{m} = \frac{\pi\ D\ h\ C_a}{v}$$

$$2.5 = \frac{\pi\ D \times 0.04 \times 30.78}{0.6686} \qquad \therefore D = 0.432\ m$$

$$U = \frac{\pi\ D\ N}{60}$$

$$N = \frac{60\ U}{\pi\ D} = \frac{60 \times 41.04}{\pi \times 0.432} = 1813.76\ rpm$$ Ans.

$$\Delta C_w = 2\,C_1 \cos\alpha_1 - U$$

$$= 2 \times 90 \cos 20^\circ - 41.04$$

$$= 169.14 - 41.04 = 128.1\ m/s$$

$$P = \frac{\dot{m}\ \Delta C_w\ U}{1000} = \frac{2.5 \times 128.1 \times 41.04}{1000} = 13.14\ kW$$ Ans.

Using cosine formula in the inlet velocity triangle

$$V_1^2 = C^2 + U^2 - 2V\ C_{a1} \cos\alpha_1$$

$$= 90^2 + 41.04^2 - 2 \times 41.04 \times 90 \cos 20^\circ$$

$$= 8100 + 1684.28 - 6941.697$$

$$= 9784.28 - 6941.697$$

$$= 2842.58$$

$$\therefore V_1 = 53.315\ m/s$$

$$\text{Energy input} = C_1^2 - \frac{V_1^2}{2}$$

$$= 90^2 - \frac{53.315^2}{2}$$

$$= 8100 - \frac{2842.58}{2} = 8100 - 1421.3 = 6678.7$$

$$\text{Diagram efficiency } \eta_d = \frac{\Delta C_w\ U}{C_1^2 - \frac{V_1^2}{2}}$$

$$= \frac{128.1 \times 41.04}{6678.7} = \frac{5257.224}{6678.7} = 0.78716 = 78.716\%$$ Ans.

Enthalpy drop in the moving blades

$$= \frac{V_2^2 - V_1^2}{2 \times 1000} = \frac{C_1^2 - V_1^2}{2000}$$

$$= \frac{90^2 - 53.315^2}{2000}$$

$$= \frac{8100 - 2842.58}{2000} = \frac{5257.42}{2000}$$

$$= 2.628\ kJ/kg$$

Therefore,

Total enthalpy drop per stage = $2 \times 2.628 = 5.257\ kJ/kg$ Ans.

2.10 At a stage of a reaction turbine the mean ring diameter is 1.5 m. The speed ratio is 0.75. The speed of rotation 50 rps. Determine the required entrance angle for the blading if the exit angle is 20°. Determine: (a) the diagram power for mass steam flow rate is 5 kg/s. (b) the diagram efficiency, (c) the enthalpy drop in moving blades (d) The percentage increase in diagram efficiency if the blades are designed for and run at the optimum operating conditions, the exit angle being kept the same. **[JNVU, 1991, 1999, 2001, 2008]**

Solution : Given

$$D = 1.5m,\ \frac{U}{C_1} = 0.75.\ N = 50\ rps,$$

$$U = \pi\ D\ N = \pi \times 1.5 \times 50 = 235.62\ m/s$$

$$C_1 = \frac{U}{0.75} = \frac{235.62}{0.75} = 314.16$$

$$\Delta\ C_w = 2\ C_1\ \cos\alpha_1 - U$$

$$= 2 \times 314.16 \cos 20^\circ - 235.62$$

$$= 590.427 - 235.62 = 354.8\ m/s$$

Using cosine formula in the inlet velocity triangle

$$V_1^2 = C_1^2 + U^2 - 2\ U\ C_1\ \cos\alpha_1$$

$$= (314.16)^2 + (235.61)^2 - 2 \times 235.62 \times 314.16\ \ \cos 20^\circ$$

$$= 98696.5 + 55516.78$$

$$= 154213.29 - 139116.567$$

$$= 15096.723$$

$$\therefore\ V_1 = 122.868\ m/s$$

Using Sine rule

$$\frac{V_1}{\sin 20^\circ} = \frac{C_1}{\sin\beta_1}$$

$$\sin\beta_1 = \frac{C_1}{V_1}\sin 20^\circ$$

$$= \frac{314.16}{122.868}\sin 20^\circ = 0.8745$$

$$\sin\beta_1 = 0.8745$$

$$\beta_1 = \sin^{-1}(0.8745)$$

$$= 60.986 = 61^\circ$$ Ans.

Energy input $= C_1^2 - \frac{V_1^2}{2} = (314.16)^2 - \frac{(122.868)^2}{2}$

$$= 98696.5 - \frac{15096.545}{2}$$

$$= 98696.5 - 7548.27 = 91148.23$$

$$P = \frac{\dot{m}\,\Delta C_w\,U}{1000} = \frac{5 \times 354.8 \times 235.62}{1000} = 417.98$$

$$= 418 \text{ kW}$$ Ans.

Diagram efficiency $\eta_d = \dfrac{\Delta C_w U}{C_1^2 - \dfrac{V_1^2}{2}}$

$$= \frac{354.8 \times 235.62}{91148.23}$$

$$= \frac{83597.976}{91148.23} = 0.917 = 91.7\%$$ Ans.

Enthalpy drop in the moving blades

$$= \frac{V_2^2 - V_1^2}{2 \times 1000} = \frac{C_1^2 - V_1^2}{2000}$$

$$= \frac{(314.16)^2 - (122.868)^2}{2000}$$

$$= \frac{98696.5 - 15096.545}{2000}$$

$$= \frac{83599.955}{2000} = 41.8\ kJ/kg$$ Ans.

$$\eta = \frac{2\cos^2\alpha_1}{1 + \cos^2\alpha_1} = \frac{2\cos^2 20^\circ}{1 + \cos^2 20^\circ} = 93.78\%$$

Percentage increase in $\eta = \dfrac{93.78 - 91.7}{91.7} \times 100$

$$= \frac{2.08}{91.7} \times 100\% = 2.268\%$$ Ans.

$$U = C_1 \cos\alpha_1 = 314.16 \cos 20^\circ = 295.21\ m/s$$ Ans.

2.11 The speed of rotation of blade group of reaction turbine is 3000 rpm. The mean blade velocity is 100 m/s. The blade speed ratio is 0.56 and the exit angle of the blades is 20°. If the mean specific volume of the steam is 0.65 m³/kg and the mean height of the blades is 25 mm, calculate the mass flow of steam through the turbine. Neglect the effect of blade thickness on the annulus area, and assume 50% reaction blading.

If five pairs of blades, calculate the useful enthalpy drop required for the group and the diagram power, calculate also the optimum diagram efficiency for the reaction stage, assuming the nozzle angle and the axial velocity remain unchanged, the blade speed being adjusted. Calculated also the blade speed at optimum efficiency. **[JNVU, 1993, 1999, 2008]**

Solution : Given

$$N = 3000\ rpm,\ U = 100\ m/s, U = 0.56\ C_1, v = 0.65\ m^3/kg, h = 25\ mm$$

$$C_1 = \frac{U}{0.56} = \frac{100}{0.56} = 178.57\ m/s$$

$$\Delta C_w = 2\ C_1 \cos \alpha_1 - U$$

$$= 2 \times 178.57 \cos 20^\circ - 100$$

$$= 335.6 - 100 = 235.6\ m/s$$

$$C_{a1} = C_1 \sin 20^\circ$$

$$= 178.57 \sin 20^\circ = 61.1\ m/s$$

$$U = \frac{\pi D N}{60}$$

$$100 = \frac{\pi D \times 3000}{60}$$

$$\therefore D = 0.636\ m$$

$$\dot{m} = \frac{\pi D h C_{a1}}{v} = \frac{\pi \times 0.636 \times 0.025 \times 61.1}{0.65}$$

$$= 4.695\ kg/s = 16903.5\ kg/hr \qquad \text{Ans.}$$

$$P = \frac{\dot{m}\ \Delta C_w\ U}{1000} = \frac{4.695 \times 235.6 \times 100}{1000} = 110.625\ kW$$

For five pairs $P = 5 \times 110.625 = 553.12\ kW$ Ans.

Using cosine formula in the inlet velocity triangle

$$V_1^2 = C_1^2 + U^2 - 2\ U\ C_1 \cos \alpha_1$$

$$= (178.57)^2 + (100)^2 - 2 \times 100 \times 178.57 \cos 20^\circ$$

$$= 31887.25 + 10000 - 33560.182$$

$$= 41887.25 - 33560.182 = 8327.06$$

$$\therefore V_1 = 91.25\ m/s$$

Enthalpy drop in moving blades

$$= \frac{V_2^2 - V_1^2}{2000} = \frac{(178.57)^2 - (91.25)^2}{2000}$$

$$= \frac{31887.25 - 8327.56}{2000} = \frac{23560.19}{2000} = 11.78\ kJ/kg$$

Therefore,

Total enthalpy drop in moving blades for 5 pairs $= 5 \times 11.78 = 58.9\ kJ/kg$

In the 50% reaction turbine

$$C_1 = V_2 \text{ and } C_2 = V_1$$

Enthalpy drop in fixed blades = 58.9 kJ/kg

Therefore total enthalpy drop $= 58.9 + 58.9 = 117.8\ kJ/kg$ **Ans.**

$$\eta = \frac{2\cos^2\alpha_1}{1+\cos^2\alpha_1} = \frac{2\cos^2 20°}{1+\cos^2 20°}$$

$$= \frac{2 \times 0.883}{1 + 0.883} = \frac{1.766}{1.883} = 0.9378 = 93.78\%$$ **Ans.**

$$\frac{U}{C_1} = \cos\alpha_1 = \cos 20° = 0.9396$$

$$\therefore U = 0.9396 \times 178.57 = 167.8\ m/s$$ **Ans.**

2.12 The first stage of a turbine is a two row velocity compounded impulse wheel. The steam velocity at inlet is 600 m/s and the mean blade velocity is 120 m/s. The nozzle angle is 16° and the exit angles for the first row of moving blades are 18°, 21° and 35° respectively. Calculate the blade inlet angles for each row. Calculate also for each row of moving blades, the driving force and the axial thrust on the wheel for a mass flow of 1 kg/s. Calculate the diagram efficiency for the wheel and the diagram power per kg/s steam flow. What would be the maximum possible diagram efficiency for the given steam inlet velocity and nozzle angle? Take the blade velocity coefficient as 0.9 for all blades. **[JNVU, 2000, 2004, 2007, 2010]**

Solution : Given

$$C_1 = 600\ m/s,\quad U = 120\ m/s,\quad \alpha_1 = 160°,\quad \beta_2 = 18°,\quad \alpha_3 = 21°$$

$$\beta_4 = 35°,\ \dot{m} = 1\ kg/s,\ K = 0.9$$

From inlet velocity triangle

$$C_{a1} = C_1 \sin\alpha_1 = 600 \sin 16° = 165.38\ m/s$$

$$C_{w1} = C_1 \cos\alpha_1 = 600 \cos 16° = 576.76\ m/s$$

$$\therefore C_{w1} - U = 576.76 - 120 = 456.76$$

$$\tan\beta_1 = \frac{C_{a1}}{C_{w1} - U} = \frac{165.38}{456.76} = 0.362$$

Inlet blade angle first row of moving blade is β_1

$$\therefore\ \beta_1 = \tan^{-1}(0.362) = 19.9^\circ$$

$$V_1 = \frac{C_{a1}}{\sin\beta_1} = \frac{165.38}{\sin 19.9^\circ} = 485.77\ m/s$$

From outlet velocity triangle

$$V_2 = K\ V_1 = 0.9 \times 485.77 = 437.1945\ m/s$$

$$C_{w2} = V_2 \cos\beta_2 - U$$

$$= 437.1945 \cos 18^\circ - 120$$

$$= 415.79 - 120 = 295.79\ m/s$$

$$C_{a2} = V_2 \sin\beta_2 = 437.1945\ \sin 18^\circ = 135.1\ m/s$$

Inlet blade angle, fixed blades is α_2

$$\tan\alpha_2 = \frac{C_{a2}}{C_{w2}} = \frac{135.1}{295.29} = 0.4567$$

$$\therefore\ \alpha_2 = \tan^{-1}(0.4567) = 24.548^\circ$$

$$C_{w2} = C_2\ .\cos\alpha_2$$

$$\therefore\ C_2 = \frac{C_{w2}}{\cos 24.54^\circ} = \frac{295.79}{0.9096} = 325.18\ m/s$$

For fixed blades: Absolute velocity at inlet to the second row of moving blades C_3 is equal to the velocity of the steam leaving the fixed row of the fixed blades.

$$C_3 = K\ C_2$$

$$= 0.9 \times 325.18 = 292.662\ \text{m/s}$$

For moving blade inlet velocity is C_3

$$C_{w3} = C_3 \cos\alpha_3 = 292.662 \cos\ 21^\circ = 273.22\ m/s$$

$$C_{a3} = C_3 \sin\alpha_3 = 292.662 \sin\ 21^\circ = 104.88\ m/s$$

$$\therefore\ C_{w3} - U = 273.22 - 120 = 153.22$$

$$\tan\beta_3 = \frac{C_{a2}}{C_{w3} - U} = \frac{104.88}{153.22} = 0.6845$$

Inlet blade angle, second row of moving blade is β_3

$$\therefore\ \beta_3 = \tan^{-1}(0.6845) = 34.39^\circ$$

$$V_3 = \frac{C_{a2}}{\sin\beta_3} = \frac{104.88}{\sin 34.39^\circ} = 185.68\ m/s$$

$$\therefore V_4 = K\ V_3 = 0.9 \times 185.68 = 167.11\ m/s$$

$$C_{w4} = V_4 \cos \beta_4 - U$$

$$= 167.11 \cos 35^\circ - 120 = 136.89 - 120 = 16.89\ m/s$$

$$C_{a4} = V_4 \sin 35^\circ = 167.11 \sin 35^\circ = 95.85\ m/s$$

$$\tan \alpha_4 = \frac{C_{a4}}{C_{w4}} = \frac{95.85}{16.89} = 5.675$$

$$\alpha_4 = \tan^{-1}(5.676) = 80^\circ$$

$$\Delta C_{aI} = C_{a1} - C_{a2} = 165.38 - 135.1 = 30.28$$

$$\Delta C_{aII} = C_{a2} - C_{a4} = 104.88 - 95.85 = 9.03$$

For first row of moving blades

$$\dot{m}\ \Delta C_{w1} = \dot{m}\ (C_{w1} + C_{w2}) = 576.76 + 295.79 = 872.55\ N$$

For second row of moving blades

$$\dot{m}\ \Delta C_{wII} = \dot{m}\ (C_{w3} + C_{w4}) = 273.22 + 16.89 = 290.11\ N$$

Total driving force

$$\therefore \dot{m}\ (C_{wI} + C_{wII}) = 872.55 + 290.11 = 1162.66\ N$$

$$P = \frac{\dot{m}\ (\Delta C_{wI} + \Delta C_{wII})\, U}{1000} = \frac{1 \times 1162.66 \times 120}{1000} = 139.52\ kW$$ **Ans.**

Axial thrust

$$F_a = \dot{m}\ (\Delta C_{aI} + \Delta C_{aII}) = (30.28 + 9.03) = 39.31\ N$$ **Ans.**

$$\eta_d = \frac{2\,(\Delta C_{wI} + \Delta C_{wII})\, U}{C_1^2}$$

$$= \frac{2 \times 1162.66 \times 120}{(600)^2} = \frac{279038.4}{360000} = 0.7751$$

$$= 77.51\%$$ **Ans.**

Maximum diagram efficiency $= \cos^2 \alpha_1 = \cos^2 16^\circ = 0.923 = 92.3\%$ **Ans.**

2.13 The nozzles of a two row velocity compounded stage have outlet angles of 22° and utilize an isentropic enthalpy drop of 200 kJ per kg of steam. All moving and guide blades are symmetrical, and the mean blade speed is 150 m/s. Assuming an isentropic efficiency for the nozzles of 90°, find all the blade angles and calculate the specific power output produced by the stage. The velocity at inlet to the stage can be neglected. **[JNVU 1996, 1999, 2005, 2009]**

Solution : Given

$$\alpha_1 = 22^\circ,\quad \Delta h = 200\ kJ/kg,\ U = 150\ m/s,\ K = 0.9$$

$$C_1 = 44.72\sqrt{k\,\Delta\,h} = 44.72\sqrt{0.9 \times 200} = 600\ m/s$$

From inlet velocity triangle

$$C_{w1} = C_1 \cos\alpha_1 = 600 \cos 22^\circ = 556.31\ m/s$$

$$C_{a1} = C_1 \sin\alpha_1 = 600 \sin 22^\circ = 224.76\ m/s$$

$$= V_1 \sin\beta_1$$

$$\tan\beta_1 = \frac{C_{a1}}{C_{w1} - U}$$

$$= \frac{224.76}{556.31 - 150} = \frac{224.76}{406.31} = 0.5532$$

$$\therefore \beta_1 = \tan^{-1}(0.5532) = 28.95^\circ$$

$$\therefore \beta_2 = \beta_1 = 28.95^\circ$$

$$V_1 \cos\beta_1 = C_{w1} - U$$

$$V_1 = \frac{406.31}{\cos 28.95^\circ} = \frac{406.31}{0.875} = 464.33\ m/s$$

Assuming $V_2 = V_1$ [Because nothing is mentioned in the question.]

From outlet velocity triangle

$$C_{a2} = V_2 \sin\beta_2 = 464.33 \sin 28.95^\circ = 224.75\ m/s$$

$$C_{w2} = V_2 \cos\beta_2 - U$$

$$= 464.33 \cos 28.95^\circ - 150$$

$$= 406.3 - 150$$

$$= 256.31\ m/s$$

$$\therefore \Delta C_{w1} = C_{\omega 1} + C_{\omega 2} = 556.31 + 256.31$$

$$= 812.62\ m/s$$

$$\tan\alpha_2 = \frac{C_{a2}}{C_{w2}} = \frac{224.757}{256.31} = 0.87689$$

$$\alpha_2 = \tan^{-1}(0.87689) = 41.247^\circ$$

$$C_2 \cos\alpha_2 = C_{w2} = 256.31$$

$$\therefore C_2 = \frac{256.31}{\cos 41.247^\circ} = 340.89\ m/s$$

For fixed blades, $C_3 = C_2$ and for moving blades $\alpha_3 = \alpha_2 = 41.247^\circ$

From inlet triangle for fixed blade

$$C_{w3} = C_3 \cos \alpha_3 = 340.89 \cos 41.247^\circ$$

$$= 256.31 \ m/s$$

$$C_{a3} = C_3 \sin \alpha_3 = 340.89^\circ \sin 41.247^\circ$$

$$= 224.76 \ m/s$$

$$\tan \beta_3 = \frac{C_{a3}}{C_{w3} - U} = \frac{224.76}{256.31 - 150} = \frac{224.76}{106.31} = 2.1142$$

$$\beta_3 = \tan^{-1}(2.1142) = 64.68^\circ$$

$$\beta_3 = \beta_4 = 64.68^\circ$$

$$V_3 \cos \beta_3 = C_{w3} - U$$

$$= 256.31 - 150 = 106.31$$

$$\therefore V_3 = \frac{106.31}{\cos 64.68^\circ} = \frac{106.31}{0.4275} = 248.634 \ m/s$$

$$V_3 = V_4 = 248.634$$

$$C_{w4} = U - V_4 \cos \beta_4$$

$$= 150 - 248.634 \cos 64.68^\circ$$

$$= 150 - 106.334 = 43.66$$

$$\Delta C_{wII} = C_{w3} + C_{w4} = 43.66 = 212.65$$

$$C_{a4} = V_4 \sin \beta_4 = 248.634 \sin 64.68^\circ = 224.75 \ m/s$$

$$\tan \alpha_4 = \frac{C_{a4}}{C_{4w2}} = \frac{224.75}{43.66} = 5.1477$$

$$\therefore \alpha_4 = \tan^{-1}(5.14770) = 79^\circ$$

$$C_4 \cos \alpha_4 = C_{w4} = 43.60$$

$$C_4 = \frac{43.66}{\cos 79^\circ} = \frac{43.66}{0.1908} = 228.82$$

$$\Delta C_w = \Delta C_{wI} + \Delta C_{wII} = 816.62 + 212.65 = 1029.27$$

$$P = \frac{\dot{m} \ \Delta C_w + U}{1000} = \frac{1 \times 1029.27 \times 150}{1000} = 154.39 \ kW$$ Ans.

2.14 Steam is supplied to a turbine at 60 bar and 450°C, and expands to 0.065 bar. The first one fifth of the overall isentropic drop is to be absorbed in two row, velocity compounded, impulse stage, the blade speed ratio being 0.22. The nozzle and guide blade outlet angles are 20°, and the blade angles at entry and exit of the first row are equal. The outlet blade angle of the second row is determined by the requirement that the axial thrust for the whole stage is to be zero. If the velocity coefficient of the nozzle is 0.98 and there is a 10% drop in relative velocity in the fixed and moving blades, find the blade speed, and the blade angles for both fixed and moving blades. **[JNVU 1986, 1990, 1995, 1998, 2000]**

Solution : Given

$$\rho = \frac{U}{C_1} = 0.22, \quad \alpha_1 = \alpha_3 = 20^\circ, \quad K = 0.9^\circ, \quad F_a = 0$$

From Mollier chart, at 60 bar 450°C and 0.065 bar.

$$\Delta h = 3295.93 - 2075.53 = 1220.46 \text{ kJ/kg}$$

As per question

$$\text{Useful drop} = \frac{\Delta h}{5} = \frac{1220}{5} = 244$$

$$C_1 = 44.72\sqrt{0.98 \times \text{useful drop}}$$

$$= 44.72\sqrt{0.98 \times 244} = 691.66 \ m/s$$

$$\therefore U = 0.22\, C_1 = 0.22 \times 691.66 = 152.16 \ m/s$$

From inlet velocity triangle

$$C_{w1} = C_1 \cos\alpha_1 = 691.66 \cos 20^\circ = 649.74 \ m/s$$

$$C_{a1} = C_1 \sin\alpha_1 = 691.66 \sin 20^\circ = 236.56 \ m/s$$

$$\therefore C_{w1} - U = 649.74 - 152.16 = 497.78$$

$$\tan\beta_1 = \frac{C_{a1}}{C_{w1} - U} = \frac{236.56}{497.78} = 0.475$$

$$\beta_1 = \tan^{-1}(0.475) = 25.418^\circ$$

$$V_1 \sin\beta_1 = C_{a1} \quad \text{and} \quad \beta_2 = \beta_1 = 25.418^\circ$$

$$\therefore V_1 = \frac{236.56}{\sin 25.410^\circ} = 551.138 \ m/s$$

$$V_2 = K\, V_1 = 0.9 \times 551.138 = 496.02 \ m/s$$

From outlet velocity triangle

$$C_{w2} = V_2 \cos\beta_2 - U = 496.02 \cos 25.48^\circ - 152.16$$

$$= 448 - 152.16 = 295.849$$

$$C_{a2} = V_2 \sin\beta_2 = 496.02 \sin 25.418^\circ$$

$$= 212.9 \ m/s$$

$$\tan\alpha_2 = \frac{C_{a2}}{C_{w2}} = \frac{212.9}{295.849} = 0.7196$$

$$\alpha_2 = \tan^{-1}(0.7196) = 35.74^\circ$$

$$\therefore C_2 = \frac{C_{w2}}{\cos\alpha_2} = \frac{295.849}{\cos 35.74^\circ} = 364.49\ m/s$$

$$\therefore C_3 = 0.9\ C_2 = 0.9 \times 364.49 = 328\ m/s$$

$$C_{w3} = C_3 \cos 20^\circ = 328 \cos 20^\circ = 308.25\ m/s$$

$$C_{a3} = C_3 \sin 20^\circ = 328 \sin 20^\circ = 121.196\ m/s$$

$$\tan\beta_3 = \frac{C_{a3}}{C_{w3} - U} = \frac{112.196}{308.25 - 152.16} = \frac{112.196}{156.1} = 0.718$$

$$\beta_3 = \tan^{-1}(0.718) = 35.7^\circ$$

$$V_3 = \frac{C_{a3}}{\sin\beta_3} = \frac{112.196}{\sin 35.7^\circ} = 192.26\ m/s$$

$$\therefore V_4 = K\ V_3 = 0.9 \times 192.26 = 173.34\ m/s$$

$$\text{Axial thrust} = 0 = \dot{m}\ [(C_{a1} - C_{a2}) + (C_{a3} - C_{a4})]$$

$$(236.96 - 212.9) + (112.196 - C_{a4}) = 0$$

$$\therefore C_{a4} = 134.856\ m/s$$

$$\sin\beta_4 = \frac{C_{a4}}{V_4} = \frac{134.856}{173.034} = 0.779$$

$$\beta_4 = \tan^{-1}(0.779) = 51.2^\circ$$

Ans.

2.15 The first stage of an impulse steam turbine is supplied with steam at 14 bar, 250°C. The velocity is compounded there being two rings of moving blades separated by a set of fixed blades. The pressure in the wheel chamber is 7 bar. Assuming 8% loss in the nozzle, find the velocity of steam as it leaves the nozzle. Given that the angle of tips of the moving blades is 30°, that the blade velocity is one fifth of that of the steam discharging from the nozzle, that the velocity of steam relative to a blade is reduced by 12% during its passage through a blade ring and that the nozzle and fixed blade angles are designed so that the steam shall pass through without shock, find the power developed and the efficiency of the stage. **[JNVU, 1988, 1990, 2008, 2010]**

Solution : Given

$$\beta_1 = \beta_2 = \beta_3 = \beta_4 = 30^\circ, \quad \frac{U}{C_1} = 0.2, \quad V_2 = 0.88\ V_1, \quad K = 0.92$$

Using Mollier chart, at 14 bar, 250°C and 7 bar

$$\Delta h = h_1 - h_2 = 2930 - 2785 = 145\ kJ/kg$$

$$C_1 = 44.72\sqrt{K\,\Delta h}$$

$$= 44.72\sqrt{0.92 \times 145} = 516.5\ m/s$$

$$\therefore U = 0.2\ C_1 = 0.2 \times 516.5 = 103.3\ m/s$$

Using Sine rule for the inlet velocity triangle

$$\frac{C_1}{\sin(180-30)} = \frac{U}{\sin(30-\alpha_1)}$$

$$\frac{516.5}{\sin 150^\circ} = \frac{103.3}{\sin(30-\alpha_1)}$$

$$\sin(30-\alpha_1) = \frac{1}{10} = 0.1$$

$$30-\alpha_1 = \sin^{-1}(0.1) = 5.739^\circ$$

$$\therefore \alpha_1 = 30 - 5.739^\circ = 24.26^\circ$$

$$C_{w1} = C_1\ \cos\alpha_1 = 516.5\cos 24.26^\circ = 470.88\ m/s$$

$$C_{a1} = C_1\ \sin\alpha_1 = 534\sin 24.26^\circ = 212.2\ m/s$$

$$V_1 = \frac{C_{a1}}{\sin\beta_1} = \frac{212.2}{\sin 30^\circ} = 424.4\ m/s$$

$$V_2 = K\ V_1 = 0.88 \times 424.4 = 373.472\ m/s$$

From outlet velocity triangle

$$C_{w2} = V_2\cos\beta_2 - U$$

$$= 373.472\ \cos 30^\circ - 103.3$$

$$= 323.436 - 103.3$$

$$= 220.136\ m/s$$

$$C_{a2} = V_2\sin\beta_2 = 373.472\ \sin 30^\circ$$

$$= 186.736\ m/s$$

$$\tan\alpha_2 = \frac{C_{a2}}{C_{w2}} = \frac{186.736}{220.136} = 0.848$$

$$\therefore \alpha_2 = \tan^{-1}(0.848) = 40.3^\circ$$

$$C_2\ \cos\alpha_2 = C_{w2}$$

$$C_2 = \frac{220.136}{\cos 40.3^\circ} = \frac{220.136}{0.7625} = 286.67$$

$$C_2 = 0.88\ C_3 = 252.27\ m/s$$

For fixed blade, using sine rule

$$\frac{C_3}{\sin(180-30)} = \frac{U}{\sin(30-\alpha_3)} = \frac{V_3}{\sin\alpha_3}$$

$$\sin(30-\alpha_3) = \frac{103.3}{252.27}\sin 150^\circ = 0.20474$$

$$30-\alpha_3 = \sin^{-1}(0.20474) = 11.814^\circ$$

$$\alpha_3 = 30 - 11.814^\circ = 18.18^\circ$$

$$C_{w3} = C_3\cos\alpha_3 = 252.27\ \cos 18.18^\circ = 239.67\ m/s$$

$$\therefore\ V_3 = \frac{C_3}{\sin 150^\circ}\ \sin 18.18^\circ = \frac{252.27\times 0.312}{0.5} = 157.418\ m/s$$

$$V_4 = 0.88\ \times 157.418 = 138.527\ m/s$$

$$C_{w4} = V_4\cos\beta_4 - U = 138.527\cos 30^\circ - 103.3$$

$$= 16.668\ m/s$$

$$\Delta C_{wI} = C_{w1} + C_{w2} = 470.88 + 220.136 = 691.016$$

$$\Delta\ C_{wII} = C_{w3} + C_{w4} = 239.67 + 16.668 = 256.338$$

$$\Delta C_w = \Delta\ C_{wI} + \Delta\ C_{wII} = 691.016 + 256.338 = 947.354$$

$$p = \frac{\dot{m}\ \Delta\ C_w \times U}{1000} = \frac{1\times 947.354\times 103.3}{1000} = 97.86\ kW$$ Ans.

Work done $= 97.86\ kJ/kg$

$$\eta_s = \frac{Work\ done}{K\ \Delta\ h} = \frac{97.86}{0.92\times 145}$$

$$= 0.7336 = 73.36\%$$ Ans.

2.16 The rows of a velocity compounded, impulse turbine have a mean blade speed of 170 m/s. The inlet and exit angles of the first moving row of blades are 30° and 25°, respectively. The inlet and exit angles of the second moving row of blades are 44° and 30°, respectively. The absolute discharge from the second row of moving blades is axial. There is a 15% loss of velocity in all blades due to friction. Determine :

(a) The inlet and exit angles of the fixed blades.

(b) The inlet nozzle angle and velocity.

(c) The power output of two rows of blades of the turbine uses 1.5 kg of steam per second. **[JNVU, 1992, 1995, 1999, 2002, 2005]**

Solution : Given

$$U = 170\ m/s,\ \beta_1 = 30^\circ,\ \beta_2 = 25^\circ,\ \beta_3 = 44^\circ$$

$$\alpha_4 = 90^\circ,\ V_2 = 0.85\ V_1,\ V_4 = 0.85\ V_3.$$

Starting from second moving blade outlet

$$V_4 = \frac{U}{\cos\beta_4} = \frac{170}{\cos 30^\circ} = 196.3\ m/s$$

$$C_4 = V_4 \sin\beta_4 = 196.3 \sin 30^\circ = 98.15\ m/s$$

$$V_3 = \frac{196.3}{0.85} = 230.94\ m/s$$

$$C_{w4} = 0$$

$$C_{a3} = V_3 \sin\beta_3 = 230.94 \sin 44^\circ$$

$$= 160.43\ m/s$$

$$C_{w3} = V_3 \cos\beta_3 + U$$

$$\therefore\ C_{w3} = 230.94 \cos 44^\circ + 170$$

$$= 166.12 + 170 = 336.12\ \ m/s$$

$$\tan\alpha_3 = \frac{C_{a3}}{C_{w3}} = \frac{160.43}{336.12} = 0.477$$

$$\alpha_3 = \tan^{-1}(0.4773) = 25.51^\circ$$

$$C_3 = \frac{C_{a3}}{\sin\alpha_3} = \frac{160.43}{\sin 25.51^\circ} = \frac{160.43}{0.43} = 372.51\ m/s$$

$$C_2 = \frac{C_3}{K} = \frac{372.51}{0.85} = 438.25\ m/s$$

Using Sine rule in the outlet velocity triangle for the first row

$$\frac{U}{\sin(\alpha_2 - \beta_2)} = \frac{C_2}{\sin\beta_2} = \frac{V_2}{\sin(180 - \alpha_2)}$$

$$\sin(\alpha_2 - \beta_2) = \frac{U}{C_2}\sin\beta_2$$

$$= \frac{170}{438.25}\sin 25^\circ = 0.1639$$

$$\alpha_2 - 25^\circ = \sin^{-1}(0.1639) = 9.4354^\circ$$

$$\therefore\ \alpha_2 = 34.435^\circ$$

$$V_2 = \frac{C_2}{\sin 25^\circ} \times \sin(180 - 34.4354)$$

$$= \frac{438.25 \sin 145.56^\circ}{\sin 25^\circ} = 586.39\ \ m/s$$

$$\therefore V_1 = \frac{V_2}{K} = \frac{586.39}{0.85} = 689.87\ m/s$$

Using Sine rule in the inlet velocity triangle for the first row

$$\frac{U}{\sin(30 - \alpha_1)} = \frac{C_1}{\sin 150^\circ} = \frac{V_1}{\sin \alpha_1}$$

$$\frac{170}{\sin(30 - \alpha_1)} = \frac{C_1}{0.5} = \frac{689.87}{\sin \alpha_1}$$

$$\therefore 689.87 \sin(30 - \alpha_1) = 170 \sin \alpha_1$$

or, $$\frac{\sin 30^\circ \cos \alpha_1 - \cos 30^\circ \sin \alpha_1}{\sin \alpha_1} = \frac{170}{689.87}$$

or, $$0.5 \cot \alpha_1 - 0.866 = 0.24642$$

or, $$\cot \alpha_1 = \frac{1.11242}{0.5} = 2.2248$$

or, $$\tan \alpha_1 = 0.44946$$

$$\therefore \alpha_1 \ \tan^{-1}(0.44946) = 24.2^\circ$$

$$\therefore C_1 = \frac{689.87 \times 0.5}{\sin 24.2^\circ} = 841.38\ m/s$$

$$C_{w1} = C_1 \cos \alpha_1 = 841.38 \cos 24.2^\circ = 767.442$$

$$C_{w2} = C_2 \cos \alpha_2 = 438.25 \cos 34.435^\circ = 361.45$$

$$\therefore \Delta C_{w1} = C_{w1} + C_{w2} = 767.442 + 361.45 = 1128.89$$

$$\Delta C_{wII} = C_{w3} + C_{w4} = 336.12 + 0 = 336.12$$

$$\therefore \Delta C_w = \Delta C_{w1} + \Delta C_{wII}$$

$$= 1128.89 + 336.12 = 1465$$

$$P = \frac{\dot{m}\, \Delta C_w\, U}{1000}$$

$$= \frac{1.5 \times 1465 \times 170}{1000} = 373.58\ kW$$ Ans.

2.17 The following particulars apply to a two row velocity compounded impulse stage of a turbine. Nozzle angle 17°, blade speed 125 m/s, exit angles of the first row moving blades, the fixed blades, and the second row moving blades, 22°, 26° and 30° respectively. Take blade velocity coefficient for each row of blades as 0.9, and assume that the absolute velocity of the steam leaving the stage is in the axial direction. Obtain

(a) Blade inlet angles for each row

(b) The diagram efficiency. **[JNVU 1985, 1988, 1997, 2007]**

Solution : Given

$$\alpha_1 = 17^\circ,\ U = 125\ m/s,\ \beta_2 = 22^\circ,\ \alpha_3 = 26^\circ,\ \beta_4 = 30^\circ,\ \alpha_4 = 90^\circ,\ K = 0.9$$

From the outlet velocity triangle for the second stage

$$U = V_4 \cos \beta_4$$

$$V_4 = \frac{125}{\cos 30^\circ} = 144.337\ m/s$$

$$V_3 = \frac{V_4}{K} = \frac{144.337}{0.9} = 160.375\ m/s$$

$$C_4 = V_4 \sin \beta_4 = 144.337 \sin 30^\circ = 72.168\ m/s$$ Ans.

Using Cosine formula in the velocity triangle of the second stage

$$V_3^2 = C_3^2 + U^2 - 2\ C_3\ U \cos \alpha_3$$

$$(160.375)^2 = C_3^2 + (125)^2 - 2\ C_3 \times 125 \cos 26^\circ$$

$$25721.74 = C_3^2 + 15625 - 224.7\ C_3$$

or, $$C_3^2 - 224.7\ C_3 - 10096.74 = 0$$

$$\therefore C_3 = \frac{224.7 + \sqrt{(224.7)^2 + 4 \times 1 \times 10096.74}}{2}$$

$$= \frac{224.7 + 301.458}{2} = \frac{526.158}{2} = 263.1\ m/s$$

$$C_{w3} = C_3 \cos \alpha_3 = 263.1 \cos 26^\circ = 236.47\ m/s$$

$$C_{a3} = C_3 \sin \alpha_3 = 263.1 \sin 26^\circ = 115.335\ m/s$$

$$\tan \beta_3 = \frac{C_{a3}}{C_{w3} - U} = \frac{115.335}{236.47 - 125} = \frac{115.335}{111.47} = 1.0346$$

$$\therefore \beta_3 = \tan^{-1}(1.0346) = 45.97^\circ$$ Ans.

$$C_2 = \frac{C_3}{0.9} = \frac{263.1}{0.9} = 292.34\ m/s$$

Using Cosine formula in the outlet velocity triangle of the first stage.

$$C_2^2 = V_2^2 + U^2 - 2\ U\ V_2 \cos \beta_2$$

$$(292.34)^2 = V_2^2 + 15625 - 2 \times V_2 \times 125 \cos 22^\circ$$

$$85458.77 = V_2^2 + 15625 - 231.795\ V_2$$

$$V_2^2 - 231.795\ V_2 - 69833.77 = 0$$

$$V_2 = \frac{231.795 + \sqrt{53728.92 + 279335.08}}{2}$$

$$= \frac{231.795 + \sqrt{333064.002}}{2}$$

$$= \frac{231.795 + 577.117}{2} = 404.45 \ m/s$$

$$V_1 = \frac{V_2}{0.9} = 449.38 \ m/s$$

$$C_{a2} = V_2 \sin \beta_2 = 404.45 \sin 22^\circ = 151.5 \ m/s$$

$$C_{w2} = V_2 \cos \beta_2 - U = 404.45 \cos 22^\circ - 125$$

$$= 375 - 125 = 250 \ m/s$$

$$\tan \alpha_2 = \frac{C_{a2}}{C_{w2}} = \frac{151.5}{250} = 0.606$$

$$\alpha_2 = \tan^{-1}(0.606) = 31.21^\circ$$ Ans.

Using Sine rule in the inlet velocity triangle for the first row of the stage ,

$$\frac{U}{\sin(\beta_1 - 17^\circ)} = \frac{V_1}{\sin 17^\circ} = \frac{C_1}{\sin(180 - \beta_1)}$$

$$\therefore \sin(\beta_1 - 17) = \frac{U \times \sin 17^\circ}{V_1} = \frac{125 \times \sin 17^\circ}{449.38} = \frac{36.546}{449.38} = 0.0813$$

$$\beta_1 - 17 = 4.6648$$

$$\beta_1 = 21.66^\circ$$ Ans.

$$C_1 = \frac{V_1 \sin 158.335}{\sin 17^\circ} = \frac{449.38}{\sin 17^\circ} \sin 158.335^\circ = \frac{165.9}{0.2923} = 567.43 \ m/s$$

$$C_{w1} = C_1 \cos \alpha_1 = 567.43 \cos 17^\circ = 542.62 \ m/s$$

$$\Delta C_{wI} = C_{w1} + C_{w2} = 542.63 + 250 = 792.636 \ m/s$$

$$\Delta C_{wII} = C_{w3} + C_{w4} = 236.47 + 0 = 236.47$$

$$\Delta C_w = 792.636 + 236.47 = 1029.1$$

$$\eta_d = \frac{2 \Delta C_w U}{C_1^2} = \frac{2 \times 1029.1 \times 1.25}{(567.43)^2}$$

$$= \frac{257276.51}{321976.8} = 0.8 = 80\%$$ Ans.

2.18 In a reaction turbine, the whirl velocity at inlet is 260 m/s and the mean blade speed is 150 m/s. The specific volume of steam at fixed blade outlet is 4.38 m³/kg and at moving blade outlet 4.8 m³/kg. The outlet angles of fixed and moving blades are 20° and 36° respectively. Assume carryover factor as 0.86 and efficiency of the nozzle is 90%. Determine :

(a) Inlet angles

(b) Total enthalpy drop in a pair of blade rings

(c) Degree of reaction,

(d) Stage efficiency. **[JNVU, 2000, 2005, 2009]**

Solution : Given

$$\alpha_1 = 20^\circ,\ \beta_2 = 36^\circ,\ C_{w1} = 260\ m/s,\ U = 150\ m/s$$

$$V_1 = 4.38\ m^3/kg,\ V_2 = 4.8\ m^3/kg,\ \phi = 0.86,\ \eta_n = 90\%$$

From the inlet velocity triangle

$$C_{w1} = C_1 \cos\alpha_1$$

$$260 = C_1 \cos 20^\circ$$

$$\therefore C_1 = \frac{260}{\cos 20^\circ} = 276.68\ m/s$$

$$\therefore C_{w1} - U = 260 - 150 = 110\ m/s$$

$$C_{a1} = C_1 \sin 20^\circ = 276.68 \sin 20^\circ = 94.63\ m/s$$

$$\tan\beta_1 = \frac{C_{a1}}{C_{w1} - U} = \frac{94.63}{110} = 0.86$$

$$\beta_1 = \tan^{-1}(0.86) = 40.7$$ Ans.

$$C_{a1} = V_1 \sin\beta_1$$

$$94.63 = V_1 \sin 40.7^\circ$$

$$\therefore V_1 = 145.1\ m/s$$

Using continuity equation and area is constant i.e. $A_1 = A_2$

$$\dot{m}_1 = \dot{m}_2$$

$$\frac{A_1 C_{a1}}{v_1} = \frac{A_2 C_{a2}}{v_2}$$

$$\therefore \frac{C_{a1}}{v_1} = \frac{C_{a2}}{v_2}$$

$$\therefore C_{a2} = \frac{94.63}{4.38} \times 4.8 = 103.7\ m/s$$

From the outlet velocity triangle

$$C_{a2} = V_2 \sin \beta_2$$

$$103.7 = V_2 \sin 36^\circ$$

$$V_2 = 176.43 \ m/s$$

$$C_{w2} = V_2 \cos \beta_2 - U$$

$$= 176.43 \cos 36^\circ - 150$$

$$= 142.736 - 150 = -7.26$$

$$\tan(180 - \alpha_2) = \frac{C_{a2}}{C_{w2}} = \frac{103.7}{7.26} = 14.276$$

$$180 - \alpha_2 = \tan^{-1}(14.276) = 85.99 = 86^\circ$$

$$\alpha_2 = 180 - 86 = 94^\circ \qquad \text{Ans.}$$

$$C_2 \sin(180 - \alpha_2) = C_{a2}$$

$$C_2 = \frac{103.7}{\sin 94^\circ} = 103.95 \ m/s$$

Enthalpy drop in fixed blades

$$= \frac{C_1^2 - \phi\, C_2^2}{2 \times 1000 \times \eta_n} = \frac{76551.82 - 0.86 \times 10806.27}{2000 \times 0.9}$$

$$= \frac{76551.82 - 9293.39}{2000 \times 0.9}$$

$$= \frac{67258.427}{2000 \times 0.9} = 37.365 \ kJ/kg \qquad \text{Ans.}$$

Enthalpy drop in moving blades

$$= \frac{V_2^2 - \phi\, V_1^2}{2 \times 1000 \times \eta_n} = \frac{31127.54 - 18106.44}{2000 \times 0.9}$$

$$= \frac{13021.09}{2000 \times 0.9} = 7.234 \ kJ/kg \qquad \text{Ans.}$$

$$\Delta h = \textit{Total enthalpy drop} = 37.365 + 7.234 = 44.6 \ kJ/kg \qquad \text{Ans.}$$

$$\textit{Degree of reaction} = \frac{\text{Enthalpy drop in moving blades}}{\text{Total enthalpy drop}}$$

$$= \frac{7.234}{44.6} = 16.22\% \qquad \text{Ans.}$$

$$\Delta C_w = C_{w1} - C_{w2} = 260 - 7.26 = 252.74$$

$$\eta_s = \frac{\Delta C_w\, U}{\Delta h} = \frac{252.74 \times 150}{44.6 \times 1000} = 0.85 = 85\% \qquad \text{Ans.}$$

EXERCISE

A. Theory

2.1 Sketch velocity diagram for a single stage impulse turbine and find the expressions for the force, work done, diagram efficiency, gross stage efficiency and axial thrust.

2.2 What is the effect of blade friction on turbine performance?

2.3 Draw a combine velocity triangle showing the effect of outlet whirl component.

2.4 Obtain an expression for maximum diagram efficiency for single stage impulse turbine.

2.5 Prove that

Stage efficiency = Diagram efficiency x Nozzle efficiency

2.6 Prove that for a single stage impulse turbine

$$\text{Blade speed ratio} = \frac{\cos \alpha_1}{2}$$

2.7 Draw a graph between diagram efficiency and blade speed ratio with an without friction losses.

2.8 Sketch and explain the effect of blade speed ratio on diagram efficiency for velocity compounded turbines.

2.9 Sketch velocity diagram of a two stage velocity compounded impulse turbine. Determine the expressions for work done, diagram efficiency gross stage efficiency and axial thrust.

2.10 Find the optimum ratio of blade speed to steam speed for a two stage velocity compounded impulse turbine. How diagram efficiency varies with blade steam velocity ratio with the increase in number of stages.

2.11 Define degree of reaction

2.12 Explain the working of a single stage reaction turbine. Sketch pressure and velocity variations along the axis of turbine. Show expansion on h-s chart.

2.13 Obtain an expression for maximum diagram efficiency for a single stage turbine.

2.14 Prove that for a single stage reaction turbine

Blade speed ratio $\rho = \cos \alpha_1$

2.15 Prove that for a single stage reaction turbine

$$\text{Maximum diagram efficiency} = \frac{2\cos^2 \alpha_1}{1 + \cos^2 \alpha_1}$$

2.16 Sketch and explain the variation of maximum diagram efficiency versus blade speed ratio for impulse and impulse reaction turbine.

2.17 What are the various losses occuring in steam turbines?

2.18 What are the advantages and disadvantages of labyrith glads used in steam turbines.

2.19 Draw neat sketches of stepped and straight through labyrinth glands.

2.20 Draw and explain various configurations of Labyrinth seals.

2.21 What methods are provided by the labyrinth glands for keeping the minimum leakage losses?

2.22 Sketch and explain the labyrinth glands used in modern steam turbines.

Section B : Unsolved problems

2.1 Show that the maximum possible efficiency of a De-lavel steam turbine is 88.3% when the nozzle angle is 20°. Deduce the formula used. [88.3%]

2.2 Calculate the maximum blade efficiency of a stage of an impulse turbine with single row wheels, assuming equiangular blades, nozzle angle = 20°, and blade velocity coefficient = 0.83.

If the blde efficiency is to be 90% of the maximum valve, what are the possible ratios of the blade speed to steam speed? Also find the corresponding blade angles. $\left[80.8\%,\ \rho = 0.32,\ \beta_1 = \beta_2 = 29°\ \ \rho = 0.618,\ \beta_1 = \beta_2 = 47°\right]$

2.3 One stage of an impulse turbine consists of a row of nozzles and one row of moving blades. The steam enters the nozzles at a pressure of 15 bar, dry saturated, with a velocity of 130 m/s. The pressure drops along the nozzle to 9 bar. The nozzles have discharge angle of 20° and the steam passes into the blades without shock. If the velocity coefficient for nozzles is 0.9, determine for maximum efficiency conditions (a) the blade angles for equiangular blades; (b) the blading efficiency, (c) the stage efficiency, [$\beta_1 = \beta_2 = 36°$, 88%, 76.3%]

2.4 A simple steam turbine receives dry and saturated steam at a pressure of 15 bar and discharges at 2 bar. The nozzle is inclined at 20° to the direction of blade motion and 10% of the available enthalpy drop is lost on friction in the nozzle. The mean diameter of the blade ring is 1 m and the rotational speed is 6000 rpm. Assuming a blade friction factor of 0.8 and taking the blades to be symmetrical, determine (i) velocity of steam at exit from the blade (ii) The diagram (blade) efficiency. [220.9 m/s, 77.4%]

2.5 A single row impulse turbine stage receives 5 kg of dry saturated steam per second at 10 bar. The steam is expanded in the nozzle to a pressure of 7 bar with an efficiency of 0.94, and discharges at an angle of 20° to the plane of rotation of the blades. Find a suitable exit angle for the blades in order that there shall be no axial thrust on the blades, allowing a blade velocity coefficient of 0.85 and a blade speed of 175 m/s. If the internal losses due to disc friction and winding amounts to 4.2 kJ/kg of steam, find the efficiency and the power of the stage. [66.5%, 223 kW]

2.6 The following data refer to one stage of an impulse steam turbine: Steam velocity = 500 m/s, blade speed = 200 m/s, Nozzle angle = 20°, Exit angle of moving blades - 25°, steam flow 5 kg/r. Neglect the effect of friction when passing through blade passages, determine (a) the inlet angle of the moving blade (b) the power, (c) the axial thrust, (d) the diagram efficiency. [32.36°, 7458.8 kW, 180 N. 89.5%]

2.7 Steam is supplied to a simple impulse turbine at 10 bar 250°C, the pressure in the wheel casing being 1.2 bar and the nozzle efficiency 91%. The steam flow rate is

5000 kg/hr. The nozzles are inclined to the plane of rotation at 20°. The blades are equiangular and have a velocity coefficient of 0.8. Assume that the ratio of mean blade speed to steam speed at nozzle exit to be 0.45, calculate (a) the blade angles (b) the power developed, and (c) the stage efficiency. [34.93°, 400 kW, 72.24%]

2.8 The nozzles of the impulse stage of a turbine receive steam at 15 bar and 300°C, and discharge it at bar. The nozzle efficiency is 95% and the nozzle angle is 20°. The blade speed is that required for maximum efficiency and the entry of steam to the blades is without shock. The blade exit angle is 5° less than the blade inlet angle. The blade velocity coefficient is 0.9 and the steam flow rate is 1350 kg/hr. Calculate (i) the diagram power in kW, and (ii) the stage efficiency. [161.55 kW, 81.7%].

2.9 Following is the data pertaining to a single row impulse turbine. Mean diameter of blade ring = 2 m, speed = 3000 rpm. Nozzle angle = 18°, Ratio of blade velocity to steam velocity = 0.5 Blade friction factor = 0.9, Blade angle at exit – 3° more than at inlet, steam supply = 30000 kg/hr. Determine (a) Power developed, (b) Diagram efficiency, (c) Steam consumption in kg/kWhr. [1382.5 kW, 84.24%, 21.7 kg, kWhr]

2.10 The following particulars refer to a single row impulse turbine :- Mean diameter of blade ring = 2.5 m, speed = 3000 rpm Nozzle angle = 20°, Ratio of blade velocity to steam velocity = 0.4. Blade friction factor = 0.8, Blade angle at exit = 3° less than that at inlet. Steam flow = 3600 kg/hr. Determine (a) Power developed, (b) blade efficiency. (c) steam consumption in kg/kWhr. [379.33 kW, 78.8%, 9.45 kg/kWhr]

2.11 Two rows of a velocity compounded impulse turbine have a mean blade speed of 150 m/s and a nozzle velocity of 675 m/s. The nozzle angle is 20°. The exit angles of the first moving row, fixed, and second row of moving blades are 25°, 25° and 30° respectively. There is a 10% loss of velocity due to friction in all blades. The steam flow is 4.5 kg/r. Determine (a) the power output of the turbine (b) the diagram efficiency. [815.74 kW, 79.57%]

2.12 Two wheels of a velocity compounded, impulse turbine have a mean blade speed of 150 m/s. The nozzle angle is 23° and the steam leaves the nozzle with a velocity of 700 m/s. The exit angles of the first moving row, fixed, and second moving row of blades are 25°, 27° and 30°, respectively. There is a 10% loss of velocity in all blades due to friction. Determine :

(a) the inlet angles of the first moving row, fixed, and second moving row of blades, (b) the power output of the two wheels/kg of steam/s (c) the diagram efficiency. [30°, 35°, 47°, 190.5 kW, 77.7%]

2.13 The following data relate to a compound impulse turbine having two rows of moving blades and one row of fixed blades in between them.

Steam velocity coming out of nozzle = 450 m/s Nozzle angle = 15°, moving blade tip angles = 30° fixed blade discharge angle = 20°. Friction loss in each blade rows = 10% of the relative velocity. Find the blade velocity, blade efficiency and specific steam consumption, for axial discharge.

$$\left[\rho = \frac{\cos \alpha_1}{2\, n} = 0.24,\ 108\ m/s,\ 79\%\ \ 45 kg/kW/hr\right]$$

2.14 In a velocity compounded impulse turbine the rotor carries two rows of moving blades and between these two rows there is a row of stationary guide blades fixed to casing. The following particulars apply to this stage.

Nozzle angle 18°, steam speed leaving nozzle 830 m/s, blade speed, 170 m/s, shockless entry of steam at blade inlets, no end thrust; exit angle of fixed blades 18°. Assume a 10% loss in relative velocity as the steam passes through each blade ring. Determine (a) all the remaing angles of the moving and fixed blades; (b) the angle at which the steam is finally discharged. (c) the carry over kinetic energy; and (d) the diagram efficiency. [β_1 =23°, β_2 = 25°, α_3 = 34°, β_3 = 29°, β_4 = 33°, (b) α_4 = 76° (c) 8.5 kJ/kg (d) 77%].

2.15 An impulse steam turbine is supplied with steam at 35 bar and 350°C, the condenser pressure being 0.05 bar. The first stage of the turbine is velocity compounded with two rings of moving blades separated by a ring of fixed guide blades. The isentropic heat drop for this stage is one quarter of that for the whole turbine. The nozzle angle is 20° and the nozzle efficiency 88%. The velocity of both rings of moving blades is one fifth of the velocity of the steam leaving the nozzle. The exit blade angles are 30° for both fixed and moving blades and the velocity coefficient for all blades is 0.9.

If the internal efficiency of the turbine is 75%, calculate the efficiency of the first stage and the percentage of the total power developed in this stage. [Stage efficiency = 66%, Percentage of power = 20%]

2.16 The first stage of an impulse turbine is compounded for velocity and has two rings of moving blades and the ring of fixed blades. The nozzle angle is 16° and the leaving angles of the blades are 20°, fixed 25°, second moving 30°. The velocity of steam leaving the nozzles is 800 m/s and the blade speed is 150 m/s. The blade velocity coefficient for first moving ring, fixed row and second moving ring may be assumed respectively as 0.85, 0.88 and 0.9. Determined the diagram efficiency, the axial thrust on each blade ring and the power developed for 4 kg/s of steam flow. [70.95%, P = 908.2 kW, Fa = 307 N.]

2.17 The first stage of a turbine is a two-row velocity compounded wheel, steam at 40 bar and 400°C is expanded in the nozzles to 15 bar and have a velocity at discharge of 700 m/s. The inlet velocity to the stage is negligible. The relevant exit angles are: nozzle 18°, first row blades 21°, fixed blades 26.5°, second row blades 35°. Take the blade velocity coefficient for all blades as 0.9. The mean diameter of the blading is 750 mm and the turbine shaft speed is 3000 rpm. Calculate; (a) The diagram efficiency, (b) the stage efficiency [70.8%, 67.4%]

2.18 The steam in a velocity compounded impulse turbine leaves the nozzle at 500 m/s. Before leaving the stage it passes through a ring of moving blades, a ring of fixed blades and another ring of moving blades. The relative steam velocity drops by 10 percent during passage through each ring of blades and the final discharge is axial. The nozzle angle is 20°, while the discharge angles are 26° for the first moving ring, 28° for the fixed ring and 30° for the second moving ring, all angles being measured relative to the plane of rotation. Assume shockless entry and determine (a) the blade speed, (b) the diagram efficiency, (c) end thrust per kg/r. [112 m/s, 79.5%, 2.5 kg]

2.19 The following particulars refer to a stage of an impulse reaction turbine, outlet angle of fixed blades = 20°, outlet angle of moving blades = 30°, Radial height of fixed blades = 100 mm, Radial height of moving blades = 100 mm, mean blade velocity = 138 m/s, Ratio of blade speed to steam speed = 0.625. Specific volume of steam at moving blade outlet = 1.235 m^3/kg, specific volume of steam at moving blade outlet = 1.305 m^3/kg. Calculate the degree of reaction, the adiabatic heat drop in pair of blade rings, and the gross stage efficiency, given the following coefficients which may be assumed to be the same in both fixed and moving blades. Carryover factor = 0.86, nozzle efficiency and Blade efficiency = 90%. [Heat drop in pair of stage = 9.369 + 24.325 = 33.694 kJ/kg, Degree of reaction = 27.8%, $\eta_{gs} = 83.55\%$]

2.20 In a stage of impulse reaction turbine, steam enters with a speed of 250 m/s at an angle of 20° in the direction of blade motion. The mean blade speed is 150 m/s when the rotor is rotating at 3000 rpm. The blade height is 100 mm the specific volumes of steam at nozzle outlet are 3.5 m^3/ kg and 4 m^3/kg respectively. The turbine develops 294 kW. Assuming the efficiency of nozzle and blades taken together as 90% and carry over coefficient as 0.8, calculate (a) the enthalpy drop in each stage, (b) degree of reaction, and (c) stage efficiency

$\lfloor \Delta h_{fb} = 29.4\ kJ/kg,\ \Delta h_{mb} = 17.346\ kJ/kg,\ \textit{Degree of reaction}\ 37.1\%,\ \eta_s = 86.4\% \rfloor$

Chapter - 3

REGENERATIVE AND REHEAT CYCLES

3.1. INTRODUCTION

All vapour power cycles must be classified as external combustion systems. The primary advantage of external combustion systems is that they generally produce less pollution and they permit the use of low grade inexpensive fuels as well as the utilization of nuclear energy.

The apparatus which enables a power cycle to be carried out is called a heat engine. Most heat engines use either a vapour or a gas as the internal energy transfer medium. The device that actually produces output work is the prime mover. A prime mover can be reciprocating piston- cylinder steam engine, a steam turbine, an internal combustion engine, and so forth.

Water is the most common working fluid used in vapour power cycles because of its many desirable characteristics, such as low cost, availability, and high enthalpy of vaporization. Other working fluids include sodium, potassium, and mercury for high temperature applications and some organic fluids such as benzene and the freons for low temperature applications.

This chapter is devoted to the discussion of steam power plants, which produce most of the electric power in the world today.

3.2 LIMITATIONS OF CARNOT CYCLE FOR VAPOURS

In addition to the inevitable irreversibilites in the compression and expansion process, the Carnot cycle is difficult and uneconomical to reproduce under practical conditions.

(1) In a actual condenser it is not easy to control the condensation process so precisely as to end up with desired quality at state 3. (Fig. 3.1).

(2) If the final quality is not correct, that is, if state 3 is not located correctly, an isentropic compression will not end at state 4. (Fig. 3.1).

(3) In a actual compressor it is difficult to compress two phase mixture (Low value of dryness fraction i.e. high moisture contents) isentropically.

(4) In the Carnot cycle the maximum temperature is remain under the critical value (for water 374.14°C). Due to this limitation thermal efficiency also lower. Any attempt to raise the maximum temperature in the cycle will involve heat

transfer to the working fluid in a single phase, which is not easy to accomplish isothermally.

(5) Work ratio is low due to irreversibilities in compression and expansion process. Due to these irreversibilities actual efficiency is low even though the ideal efficiency is high.

(6) If superheated steam is used the Carnot cycle would be still more difficult to realise as heat of superheat would have to be supplied at constant temperature, whereas in actual practice it is only possible at constant pressure.

(7) The quality of the steam dryness fraction decreases in the isentropic expansion process from 1 to 2. The liquid water droplets in high velocity exhaust steam cause pitting, and hence erosion of the turbine blades. Thus steam with qualities less than about 90 per cent ($x_2 \approx 0.9$) cannot be tolerated in the operation of power plants. This problem could be eliminated by using a working fluid with a very steep saturated vapour line.

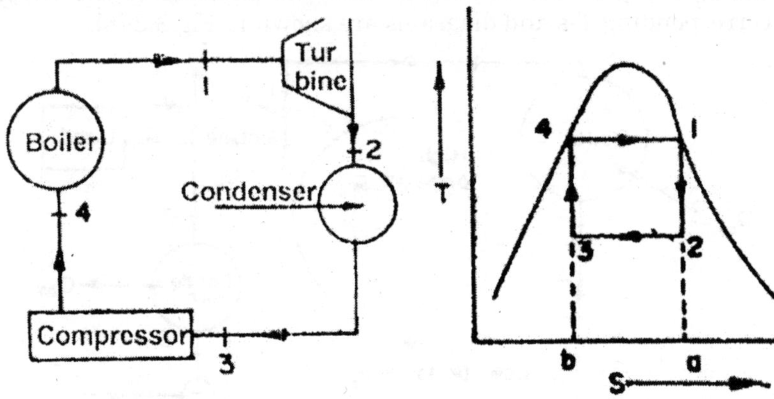

Fig. 3.1. Carnot cycle (a) Diagram matic sketch (b) T-s diagram

The Carnot cycle eliminates the difficulties of the first one but presents some new ones. The isentropic compression 3-4 involves extremely high pressures. Also, the isothermal heat addition process involves a pressure variation that introduces a control problem not encountered when the temperature can be held constant simply by maintaining a constant pressure as in heat addition to a two phase mixture.

In summary, the Carnot cycle is not suitable for the design of steam power plants.

Theoretically the Carnot cycle is the most efficient cycle but it is not taken as a standard of reference for vapour power plants.

3.3 RANKINE CYCLE

The modification made to the Carnot cycle to produce the Rankine cycle is that instead of stopping the condensation in the condenser at some intermediate section, the condensation is completed. At this point there is all water. This water can be pumped easily with a small pump in which its pressure can be raised in feeding it back into the boiler. Then in the place of bulky and expensive compressor we need only a small pump. With the use of small pump work ratio is increased and steam consumption is reduced.

Since the Rankine cycle is characterized by constant pressure heating, there is no reason to stop heating the steam when it reaches the saturated vapour state. The customary practice is to superheat the steam to a much higher temperature, because without superheat, the steam is wet throughout the expansion, which leads to excessive wall condensation in a reciprocating engine and to erosion by water droplets of the blades in a turbine.

In summary, many of the impracticalities associated with the Carnot cycle can be eliminated by superheating the steam in the boiler and condensing it completely in the condenser. The cycle that results is the Rankine cycle, which is the ideal cycle for vapour power plants. The ideal Rankine cylce does not involve any internal irreversibilities.

An ideal simple Rankine cycle is shown in Fig 3.2. Fig. 3.2 (a) is a flow diagram showing the boiler producing vapour at constant pressure through the addition of heat, an ideal turbine (isentropic expansion) producing work, a condenser having no pressure drop in which heat is removed and the vapour is condensed, and an ideal pump (isentropic) raising the pressure of the liquid to the pressure of the vapour generator (boiler). The corresponding T-s and diagrams are shown in Fig 3.2 (b).

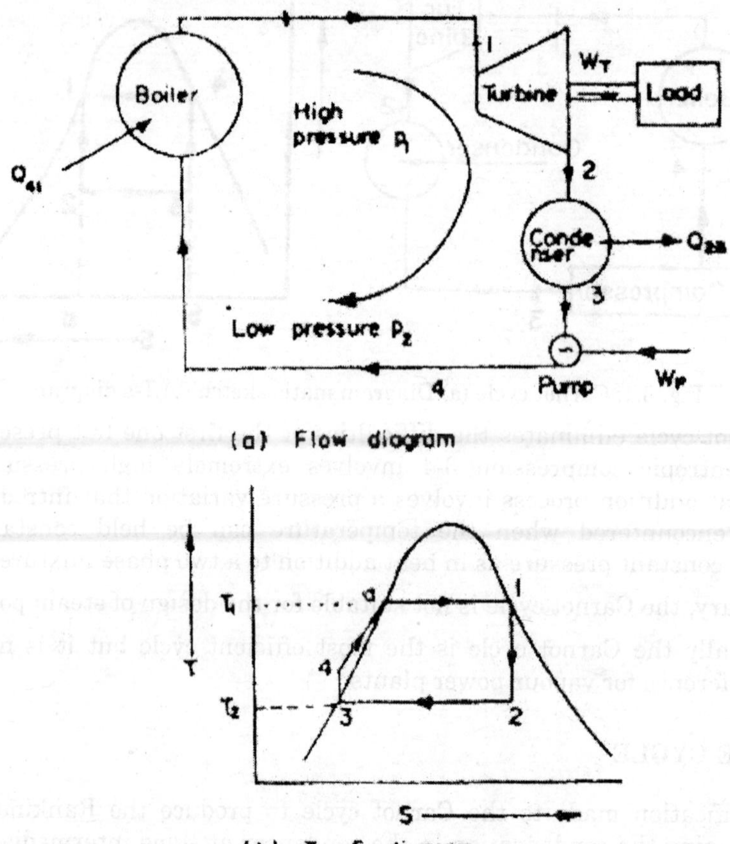

Fig. 3.2 Rankine cycle with dry and saturated.

Referring to Fig. 3.2, the working fluid undergoes the following series of internally reversible processes.

Process 1-2: Dry saturated working fluid (steam) expands isentropically to pressure p_2 or condenser pressure.

Process 2-3: The steam is then condensed at constant pressure and temperature to a saturated liquid, state 3. In the condenser, heat is transferred from the condensing steam to water.

Process 3-4: The saturated liquid leaving the condenser is then pumped isentropically into the boiler at pressure p_1 (State 4).

Process 4-1: Heat transfer to the working fluid as it flows at constant pressure p_1 through the boiler to complete the cycle.

In order to use higher temperatures without increasing the maximum pressure of the cycle, the steam after having the boiler is heated further at constant pressure in a superheater. The purpose of superheating is to avoid a large percentage of moisture in the steam during the later stages of expansion in the turbine. For the given pressure limits the thermal efficiency is also increased by superheating. T-s, diagram of a Rankine cycle with superheat is shown in Fig. 3.3.

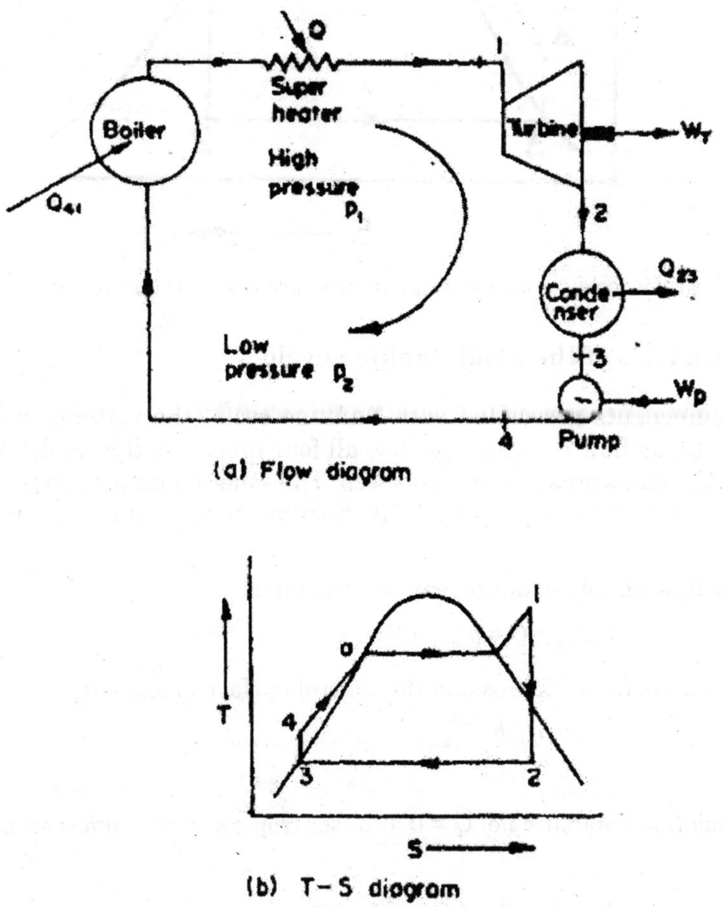

Fig. 3.3 Rankine cycle with superheat

In most actual Rankine cycle systems the steam is supplied to the turbine in the superheated state.

If the steam generator pressure is higher than the critical pressure, there can be no constant heat addition process in a Rankine cycle. T-s, p-v and h-s diagrams for such a cycle are shown in Fig 3.4.

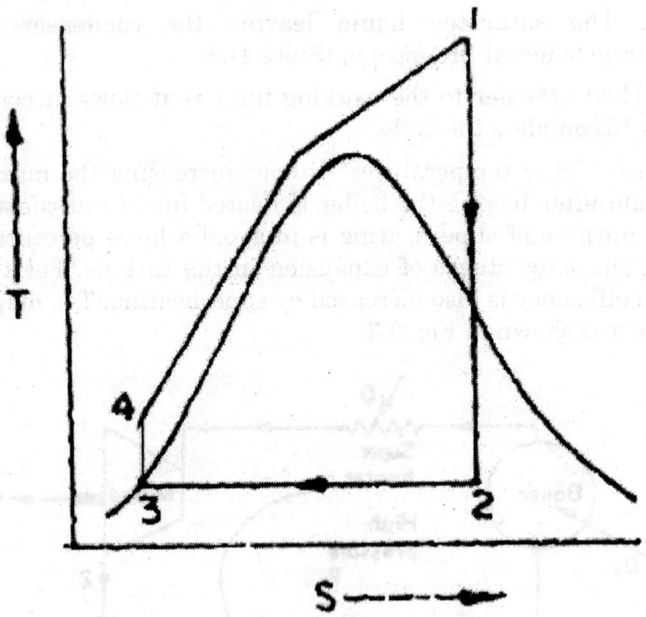

Fig. 3.4 Rankine cycle – steam generation pressure higher than the critical pressure

3.3.1 Energy analysis of the ideal Rankine cycle

All four components associated with Rankine cycle (the turbine, condenser, pump and boiler) are steady flow devices, and thus all four processes that make up the Rankine cycle can be analyzed as steady flow processes. The kinetic and potential energy changes are small relative to the work and heat transfer terms and are therefore usually neglected.

The steady flow energy equation can be written as

$$h_{initial} + Q = h_{final} + W$$

Boiler : There is no work crossing the control surfaces i.e. $w = 0$

$$Q_{14} = h_1 - h_4 \tag{3.1}$$

Turbine:

The expansion is adiabatic i.e. $Q = 0$ and isentropic $s_1 = s_2$ and h_2 can be calculated.

Then

$$h_1 = h_2 + W_{12} + Q_{12}$$

Therefore, $$W_{12} = (h_1 - h_2) \tag{3.2}$$

Condenser : The assumption used for the boiler are also applied to the condenser i.e W=0

$$Q_{23} = h_2 - h_3 \tag{3.3}$$

Pump : The pump is adiabatic i.e. $Q = 0$ and reversible $s_3 = s_4$

Then

$$h_3 + W_{34} = h_4 + Q_{34}$$

Therefore, $$W_{34} = h_4 - h_3 \tag{3.4}$$

It can be shown that for a liquid, which is incompressible (i.e v = constant). The increase in enthalpy for isentropic compression is given by

$$h_4 - h_3 = v(p_4 - p_3)$$

Where, v can be taken from steam tables for water at the pressure p_2.

Therefore, $v = v_{f2}$

Here $p_4 = p_1$ and $p_3 = p_2$

Therefore

$$W_{34} = h_4 - h_3 = v(p_4 - p_3) = v_{f2}(p_1 - p_2) \tag{3.5}$$

This is the feed pump term, and it is a small quantity in comparison with the turbine work, W_{12} it is usually neglected, especially when boiler pressure are low.

Net work done in the cycle $W_{net} = W_T - W_p$

$= W_{12} - W_{34} = (h_1 - h_2) - (h_4 - h_3)$

The heat supplied in the boiler $Q_{41} = h_1 - h_4$

$$\text{Rankine efficiency} = \frac{\text{Network output}}{\text{Heat supplied in the boiler}}$$

$$\eta_R = \frac{(h_1 - h_2) - (h_4 - h_3)}{(h_1 - h_4)} = \frac{(h_1 - h_2) - (h_4 - h_3)}{(h_1 - h_3) - (h_4 - h_3)} \tag{3.6}$$

The equation (3.6) can be written as

$$\eta_R = \frac{W_T - W_p}{(h_1 - h_3) - W_p} = \frac{(h_1 - h_2) - W_p}{(h_1 - h_3) - W_p} \tag{3.7}$$

If the feed pump term is neglected, equation (3.6) or (3.7) becomes

$$\eta_R = \frac{h_1 - h_2}{h_1 - h_3} \tag{3.8}$$

As per the ideal Rankine cycle condition, h_3 is always available at a low pressure p_2, i.e. at completely condensed condition ($h_3 = h_{f2}$). Therefore, the equation (3.8) reduces to

$$\eta_R = \frac{h_1 - h_2}{h_1 - h_{f2}} \tag{3.9}$$

3.3.2 Actual Rankine cycle

The operation of actual vapour cycle can deviate from the performance of the ideal cycle primarily because or irreversibilities associated with mechanical and fluid friction as well as additional losses.

Fig. 3.5 illustrates typical deviations that can occur between the ideal Rankine cycle and the actual cycle.

Fluid friction causes pressure drops in the boiler, the condenser, and the piping between various components. As a result, steam leaves the boiler at a somewhat lower pressure. Also the pressure at the turbine inlet is some what lower than that at the boiler exit due to the pressure drop in the connecting pipes. The pressure drop in the condenser is usually very small. To compensate for these pressure drops, the water must be pumped to a sufficiently higher pressure than the ideal cycle calls for. This requires a larger pump and larger work input to the pump.

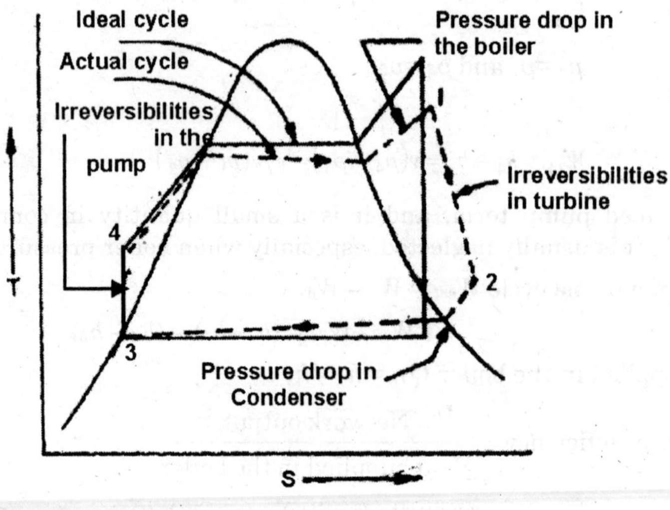

Fig. 3.5 Deviation of actual vapour cycle from the ideal Rankine cycle.

Heat transfer, which takes place from the water in the turbine to the surroundings, is another source of irreversibility. As a result of unavoidable irreversibilities that occur in the turbine, an actual turbine is capable of providing less power than an isentropic turbine. Due to these losses cycle efficiency decreases. To maintain the same level of net work output, more heat needs to be transferred to the steam in the boiler to compensate for these undesired heat losses. Pump losses are similar to those occuring in the turbine. They are due to non isentropic compression, but are often neglected because they are relatively small.

3.3.3. The effect of operating conditions on Rankine cycle efficiency

The efficiency of the Rankine cycle may be improved by either increasing the average level of temperature at which heat is supplied, or by reducing the temperature at which heat is rejected. This principle may be implemented by a suitable alteration in the conditions of steam generation or condensation (Fig. 3.6).

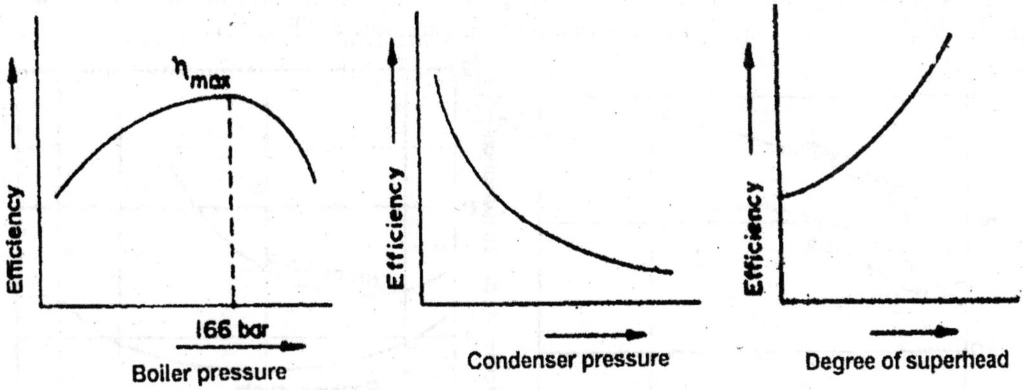

Fig. 3.6. Effect of operating conditions on the thermal efficiency of the Rankine cycle.

(a) Increasing boiler pressure

If we suppose that the steam entering the expander remains, say, dry and saturated, then an increase in boiler pressure will represent an increase in saturation temperature and on this account cycle efficiency will tend to rise. It should be realized however that this effect tends to be offset by a reduction in the magnitude of the latent heat of steam at high pressures. This results in a larger proportion of the net heat being supplied to the feed water in elevating it to saturation temperature. The overall consequence of these conflicting factors is to cause the efficiency to reach a maximum value at a boiler pressure of about 166 bar (Fig. 3.6).

(b) Superheating

Assuming all other factors remain the same, the cycle efficiency may be increased by heating the steam at constant pressure in a super heater before allowing it to expand. A collateral advantage is found in the substantial transference of the expansion process to the superheated vapour field. Longer turbine blade life then ensues from the reduction, or absence, or erosion from the high velocity water particles that are suspended in wet vapour (Fig. 3.6).

(c) Reducing condenser pressure

A reduction in the condenser pressure and hence a reduction in temperature of heat rejection has a marked effect in improving thermal efficiency, especially in high vacuums. However, the cost of the condensation apparatus may well outweigh the thermodynamic gain from increasing the vacuum (Fig. 3.6).

3.4 COMPARISON OF RANKINE AND CARNOT CYCLES

Rankine cycle provides a higher specific work output than a Carnot cycle operating between the same temperature limits and, therefore, requires a smaller steam flow rate resulting in smaller size plant for a given power output. The Rankine cycle, however, calls for higher rates of heat transfer in boiler and condenser (Fig. 3.7).

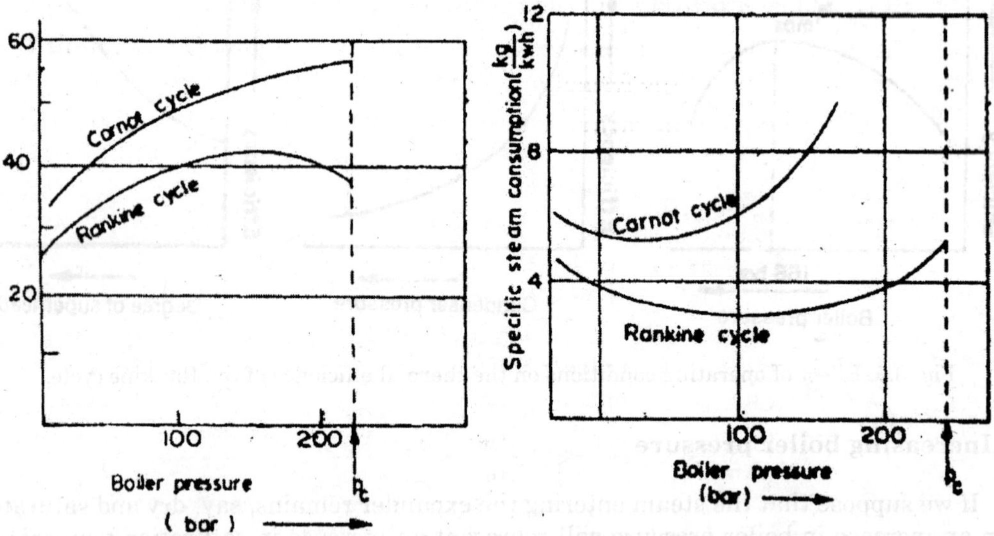

Fig. 3.7 Efficiency and specific steam consumption of Carnot and ideal Rankine cycles.

The advantage of using pump to feed liquid to the boiler instead to compressing a wet vapour is obvious that the work for compression is very large compared to pump.

In the more practical of the two cycles, the Rankine cycle, the effect of irreversibility in the pump on the cycle efficiency is negligible since the pump work is so small in comparison with the turbine output. Thus the major irreversibility to be taken into account is that occurring in the turbine.

Rankine cycle will have lower efficiency than the Carnot cycle because in Rankine cycle only part of heat is supplied isothermally at constant higher temperature T_1 (Fig. 3.7).

The efficiency of the Rankine cycle will approach that of the Carnot cycle more nearly if the superheat temperature rise is reduced. If superheat temperature rise is reduced, the steam is wet throughout the expansion, which leads to erosion by water droplets of the blades in a turbine, hence shorter turbine blade life. Due to fixed upper temperature, the pressure of the steam will rise to values which are difficult to deal with structurally if the superheat is reduced (Fig. 3.7).

3.5 DESIRABLE CHARACTERISTICS OF A GOOD WORKING FLUID

The ideal working fluid for a heat engine should make the Rankine cycle as close to the Carnot cycle as possible. More specifically, the ideal working fluid should have the following characteristics (Fig. 3.8).

(1) A working fluid must have **chemical stability**, and it should not have a corrosive effect on the metals ordinarily employed in machineries.

(2) **High density** at operating temperature and pressure this would minimize the size of the equipment.

(3) It should be **nontoxic** and **inexpensive.**

(4) It should have a **triple point temperature well below the ambient temperature** to prevent the formation of solids (i.e. freezing) within the system.

(5) **The critical temperature should be well above the metallurgical limit of the boiler and turbine** so that efficient isothermal high temperature heat transfer can occur in the boiler.

(6) It should have the **slope of its saturated vapour and liquid lines as nearly vertical as possible** on T-s diagram to avoid unnecessary superheating and reheating. (Fig. 3.8).

(7) It should have **large phase change enthalpies** (h_{fg}) **and low liquid specific heats** so that the heat required to bring the liquid condensate up to the vaporization temperature is a small percentage of the vaporization heat.

(8) The pressure temperature relation should be such that moderate pressure (reasonable pressure from the construction point of view) can be obtained **at high temperatures, and the vapour pressure at condensing temperatures** (near to ambient temperatures) should be nearly atmospheric so that excessive power for maintaining vacuum will not be required. The critical temperature should be well above any desirable upper limit of temperature of a cycle.

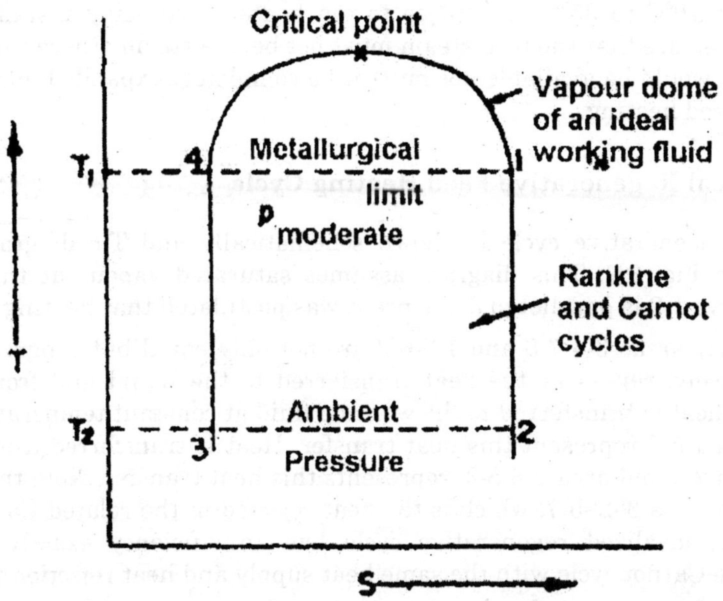

Fig. 3.8. The Rankine cycle with ideal working fluid becomes a Carnot cycle

Of course, no substance could satisfy all the above requirements. The fluids adopted or suggested for vapour cycle power plants have been steam, mercury, diphenyl oxide, zinc ammonia chloride, and fluoriented hydrocarbons (Freon compounds).

3.6 REGENERATIVE FEED HEATING

Regenerative feed heating cycle is the modified form of Rankine cycle with an aim to achieve higher thermal efficiency. It is a practical approach towards approaching Carnot cycle based on the philosophy of increasing average temperature at which heat is added to the cycle. All the steam power plants as single unit producing 30 MW and above employ regenerative feed heating arrangement.

3.6.1 Basic Principle

The aim of regenerative feed heating cycle is to supply the feed water to the boiler at a temperature much higher than the temperature of the condensate in the condenser by heating the feed water in one or more feed heaters with the help of steam bled off during expansion from steam turbine thereby increasing the average temperature of heat addition in the cycle which in turn increases the thermal efficiecy of regenerative feed heating cycle.

The feed water is heated either in a *direct contact* feed heater or in a surface (indirect) heater due to condensation of extracted steam i.e. the heat is given up by the extracted steam which is received by the feed water. In this way, the average temperature of heat addition to the cycle is increased, which increases the thermal efficiency of the cycle. This type of system has become a standard practice in steam power plants both on land and, in the more powerful installations. *In practice*, the appreciable cost of heaters, pipes, valves and accessories put a limit to the number of heaters employed. In large units, the temperature rise of feed water per heater is normally about 20°C to 35°C, and three to ten heaters are being installed. The main points to be noted are that the bled steam must not be live steam otherwise no work from that live steam would be available nor must it be completely expanded, for then it would be useless for feed heating.

3.6.2. Most Ideal Regenerative Feed Heating Cycle

An ideal regenerative cycle is shown schematically and T-s diagram (neglecting pump work) in Fig. 3.9. This diagram assumes saturated vapour at the initiation of expansion. Curve 1-2 is parallel to 3-4, since it was postulated that heating is reversible.

Consquently areas 3-4-7-8 and 1-5-6-2 are not only equal but congruous and these areas, respectively, represent the heat transferred to the liquid and from the vapour. Note also that heat is transferred to the working fluid at constant temperature in process 4-1, and area 4-1-5-7 represent this heat transfer. Heat is transferred from the working fluid in process 2-3 and area 2-6-8-3, represents this heat transfer. Note that this area is exactly equal to area 3'-2'-5-7, which is the heat rejected in the related Carnot cycle 3'-4-1-2'. Thus this idealized regenerative cycle has an efficiency exactly equal to the efficiency of the Carnot cycle with the same heat supply and heat rejection temperatures.

However theoretically admirable this scheme may appear, its practical interpretation is beset by difficulties.

(1) If the steam expanding through the turbine is cooled in order that the feed water temperature may be raised, then it becomes unduly wet. In other words

the steam expanding through the turbine would have an impracticable low dryness fraction that causes erosion of turbine blades.

(2) The turbine cannot be designed to act simultaneously as a steam expander and as an effective contra flow heat exchanger.

The above two considerations lead to a more realistic approach to regeneration.

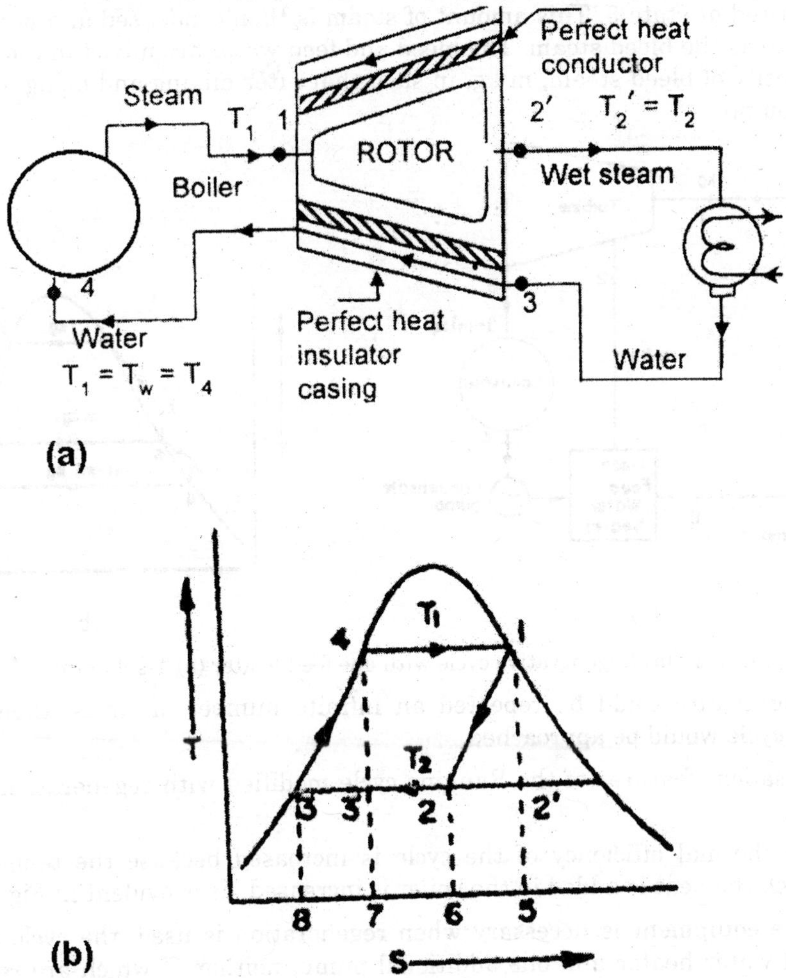

Fig. 3.9 (a) Ideal regenerative cycle, (b) T-s diagram

3.6.3 Steam bleding

A better solution is to extract or "bled" steam from the turbine at various points in the expansion process and to use this bleed steam to heat water flowing from the pump to the boiler. This energy transfer occurs in feed water heaters.

In the steam bleding system the tapped steam heats the feed water at the expense of being completely condensed at the pressure of its extraction, whilst the main bulk of the steam continues to expand adiabatically. Consequently no blade erosion problem is encountered due to steam cooling by heat transfer.

A simple regenerative system with single stage feed water heating is shown in Fig. 3.10. Consider a unit mass of steam upon its entrance to the turbine (state 1). After expansion, some of steam say m *kg* is extracted and enters the feed water heater where it is used to heat the condensate. The rest of the steam (1–*m*) *kg*, completes the expansion and is exhausted at state 3. This amount of steam is then condensed and pumped to the same pressure as the bleed steam. The bleed and feed water are mixed in the feed water, and the quantity of bleed steam, m kg, in such that after mixing and being pumped in a second feed pump.

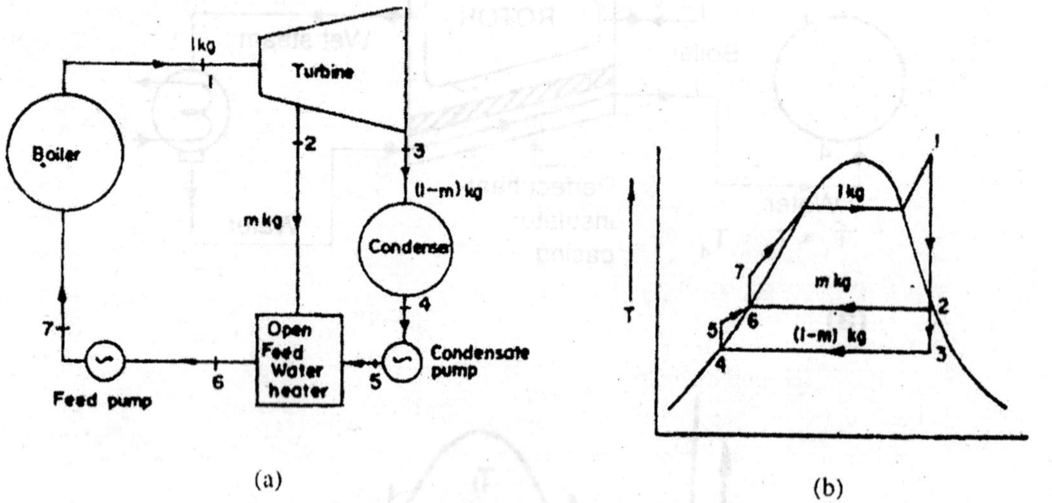

Fig. 3.10 (a) Regenerative cycle with one feed heater (b) T-s diagram

If this procedure could be repeated an infinite number of times, then the ideal regenerative cycle would be approached.

Several salient features of the Rankine cycle modified with regeneration should be recognised.

(1) The thermal efficiency of the cycle is increased because the temperature at which the heat is added in the boiler is increased, as is evident in Fig. 3.10.

(2) More equipment is necessary when regeneration is used; the cycle requires a feed water heater and one additional pump, neither of which are required for the unmodified Rankine cycle.

3.6.4. Regenerative efficiency

In Figures 3.10 (a) and 3.10 (b), the symbol m is the fraction of the total mass in the system, which leaves the turbine to go to the feedwater heater.

$$m = \frac{\text{Mass of steam feed water}}{\text{Total system mass}}$$

Now, We perform an energy balance on the feed water heater. From Fig. 3.11.

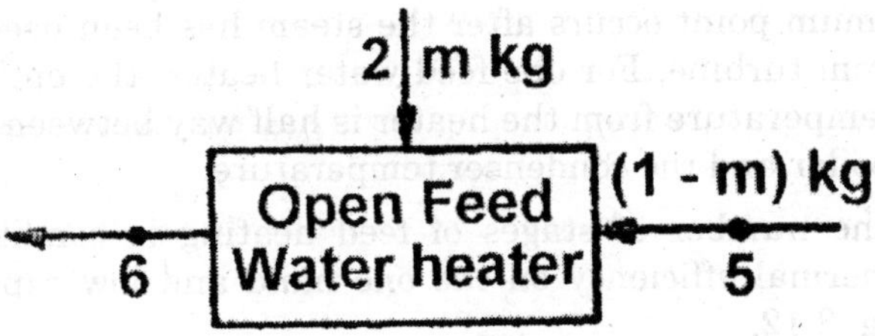

Fig. 3.11 Feed water heater – energy balance

Energy in = Energy out

$$mh_2 + (1-m)h_5 = h_6$$

$$mh_2 + h_5 - mh_5 = h_6$$

$$m(h_2 - h_5) = h_6 - h_5$$

$$m = \frac{h_6 - h_5}{h_2 - h_5} \quad (3.10)$$

When the turbine work and pump work are calculated, attention must be paid to the fraction of fluid passing through each device. The turbine work per kg in the system is.

$$W_T = (h_1 - h_2) + (1-m)(h_2 - h_3) \ kJ/kg \quad (11.27)$$

The pump work per kg in the system will have to be modified by the fraction of the total mass that passed through it. There are two pumps, a condensate pump and a feed pump.

$$W_{P1} = (h_5 - h_4)(1-m)$$

and $$W_{P2} = (h_1 - h_6)$$

Therefore, $$W_P = W_{P1} + W_{P2}$$

$$= (h_5 - h_4)(1-m) + (h_1 - h_6) \quad (3.11)$$

The heat abstracted in the condenser is

$$W_C = (1-m)(h_3 - h_4) \quad (3.12)$$

Equations (3.12) to (3.13) enable the thermal efficiency of the regenerative cycle to be obtained. Neglecting pump work.

$$\eta = \frac{\text{Heat supplied } - \text{ Heat rejected}}{\text{Heat supplied}} = \frac{W_T - W_P}{W_T}$$

$$= \frac{(h_1 - h_2) + (1-m)(h_2 - h_3) - (1-m)(h_3 - h_4)}{(h_1 - h_2) + (1-m)(h_2 - h_3)} \quad (3.14)$$

3.7 OPTIMUM HEATER LOCATION AND NUMBER OF HEATER USED

The optimum point for a feed heater lies between the turbine inlet and exit. The determination of the location is a trial and error procedure. It is sufficient to present to realize that the optimum point occurs after the steam has been used for some work and before it exhausts from turbine. For one feed water heater, the optimum point is where the feed water exit temperature from the heater is half way between the saturated steam temperature in the boiler and the condenser temperature.

The choice of the number of stages of feed heating is conditioned by conflicting desires for a high thermal efficiency on the one hand and low capital and maitenance costs on the other Fig. 3.12.

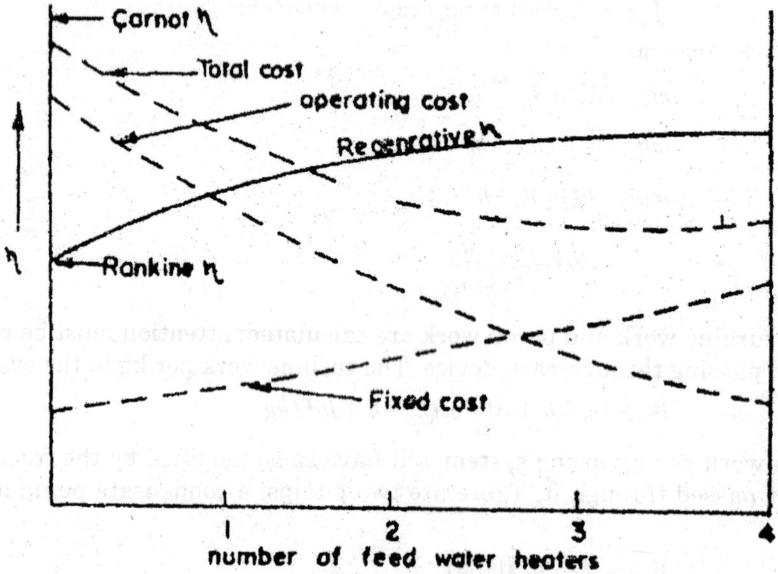

Fig. 3.12 Otimum number of feed water heates for regenerative cycle

Fig. 3.13 shows the percent decrease in cycle heat rate (corresponding to increase in cycle efficiency) For 1, 2, 3, 4 and 10 feed water heaters versus the total rise above the condenser temperature.

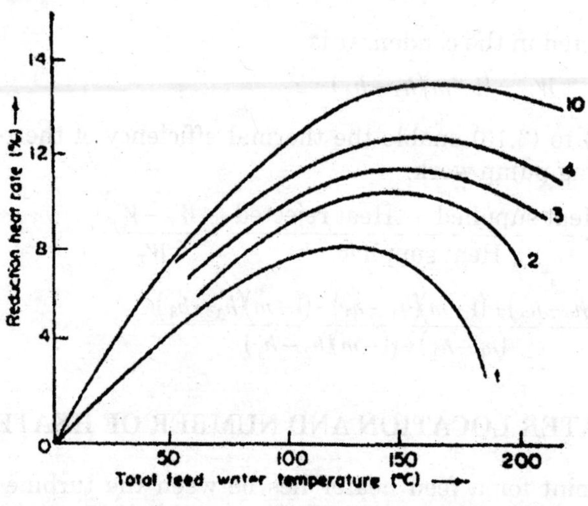

Fig. 3.13 Graph between total feed water temperature and percentage heat rate

In fact, more than seven stages are rarely used and frequently only one or two feed heaters are fitted. Many large plants in operation today use as many as eight feed water heaters. The optimum number of feedwater heaters is determined from economical considerations. The use of an additional feedwater heater cannot be justified unless it saves more from the fuel costs than its own cost.

3.8 OTHER TYPE REGENERATVE FEED HEATING ARRANGEMENTS.

3.8.1 Surface Heaters with one Drain Pump

Fig. 3.14 shows an another type of feed heating system called regenerative feed heating system with one drain pump. In this case also the same assumptions are valid.

The drain water is cascaded from heater to heater and sent to the succeeding heating at low pressure side through the steam traps T. The whole of the condensed bled steam is extracted from heater No. 4 by a drain pump and discharged to the main feed on the outlet side of the heater No. 4, on high pressure side.

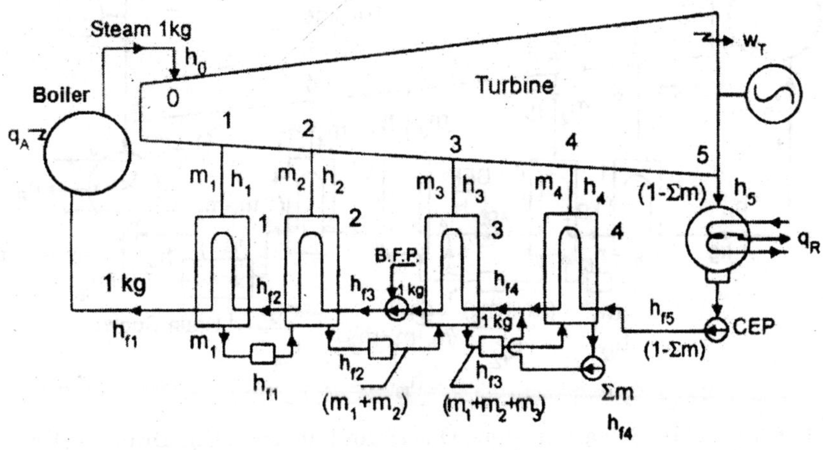

Fig. 3.14. Regenerative Feed Heating System with Four Plain Surface Heaters and one Drain Pump.

The various amount of steam extracted are :-

At Heater No. 1 : $m_1 h_1 + h_{f2} = h_{f1} + m_1 h_{f1}$

or $$m_1 = \frac{h_{f1} - h_{f2}}{h_1 - h_{f1}} \tag{3.15}$$

At Heater No. 2 : $m_2 h_2 + h_{f3} + m_1 h_{f1} = (m_1 + m_2) h_{f2} + h_{f3}$

or $$m_2 = \frac{(h_{f2} - h_{f3}) - m_1 (h_{f1} - h_{f2})}{(h_2 - h_{f2})} \tag{3.16}$$

At Heater No. 3 : $m_3 h_3 + (m_1 + m_2) h_{f2} + h_{f4} = (m_1 + m_2 + m_3) h_{f3} + h_{f3}$

or $$m_3 = \frac{(h_{f3} - h_{f4}) - (m_1 + m_2)(h_{f2} - h_{f3})}{h_3 - h_{f3}} \quad (3.17)$$

At Heater No. 4 : $m_4 h_4 + (1 - m_1 - m_2 - m_3 - m_4) h_{f5} + (m_1 + m_2 + m_3) h_{f3} = h_{f4}$

or $$m_4 = \frac{(h_{f4} - h_{f5}) - (m_1 - m_2 - m_3)(h_{f3} - h_{fw5})}{h_4 - h_{f5}} \quad (3.18)$$

The expression thermal efficiency will be on the line as discussed above

3.8.2 Plain Surface Heaters with One Drain Pump and One Drain Cooler

Fig. 3.15 shows an alternative system which is called regenerative feed heating with one drain pump and one drain cooler.

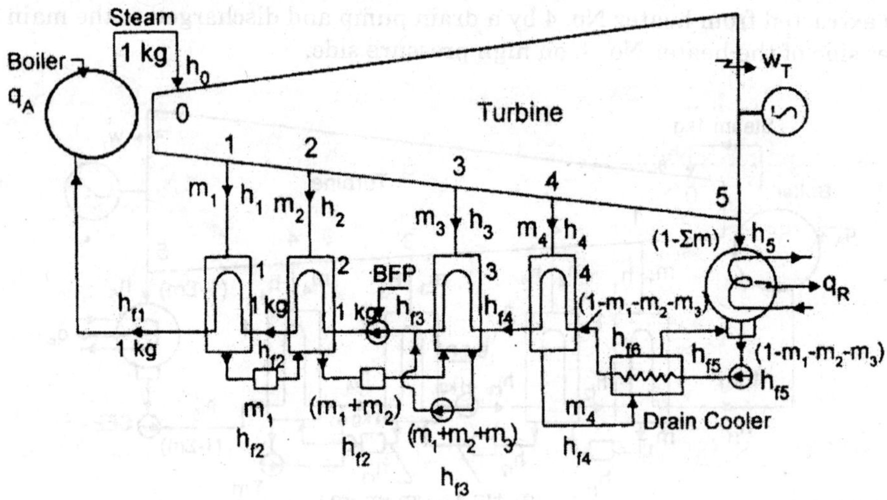

Fig. 3.15. Plain Surface Heaters, One Drain Pump and One Drain Cooler.

The condensed heating steam m_1 is cascaded from heater No. 1 to heater No. 2. From the heater No. 2 the condensed steam $(m_1 + m_2)$ is cascaded to heater No. 3. From the heater No. 3 the condensed steam $(m_1 + m_2 + m_3)$ is extracted by a drain pump and delivered to the main feed pipe immediately after the heater i.e. on the delivery side of the heater No. 3.

The drain from heater No. 4 is passed through a drain cooler to transfer some of the heat to the feed water and after cooling to condenser temperature it goes to the lower part of the condenser or hot-well.

At Heater No. 1 : $m_1 h_1 + h_{f2} = h_{f1} + m_1 h_{f1}$

or $$m_1 = \frac{h_{f1} - h_{f2}}{h_1 - h_{f1}} \quad (3.19)$$

At Heater No. 2 : $m_2 h_2 + h_{f3} + m_1 h_{f1} = (m_1 + m_2) h_{f2} + h_{f2}$

or $$m_2 = \frac{(h_{f2} - h_{f3}) - m_1 (h_{f1} - h_{f2})}{h_2 - h_{f2}} \quad (3.20)$$

At Heater No. 3 : $$m_3 h_3 + (m_1 + m_2) h_{f2} + (1 - m_1 - m_2 - m_3) h_{f4} = h_{f3}$$

or $$m_3 = \frac{(h_{f3} - h_{f4}) - (m_1 + m_2)(h_{f2} - h_{f4})}{h_3 - h_{f4}} \quad (3.21)$$

At Heater No. 4 : $$m_4 h_4 + (1 - m_1 - m_2 - m_3) h_{f6} = (1 - m_1 - m_2 - m_3) h_{f4} + m_4 h_{f4}$$

or $$m_4 = \frac{(1 - m_1 - m_2 - m_3)(h_{f4} - h_{f6})}{h_4 - h_{f4}} \quad (3.22)$$

At Drain Cooler : $$(1 - m_1 - m_2 - m_3) h_{f5} + m_4 h_{f4} = (1 - m_1 - m_2 - m_3) h_{f6} + m_4 h_{f5}$$

$$h_{f6} = \frac{(1 - m_1 - m_2 - m_3) h_{f5} + m_4 (h_{f4} - h_{f5})}{(1 - m_1 - m_2 - m_3)} \quad (3.23)$$

After getting the value of h_{f6}, m_4 can be calculated.

The expression for thermal efficiency with be on the line as discussed above.

3.8.3 Surface Heaters with Drain Fed to Hot-Well

Fig. 3.16 shows yet another type of feed heating system. In this, a hot-well which in an integeal part of condenser receives the drain from heaters. Here, only two heaters are shown and the assumptions are the same here as in the other ones.

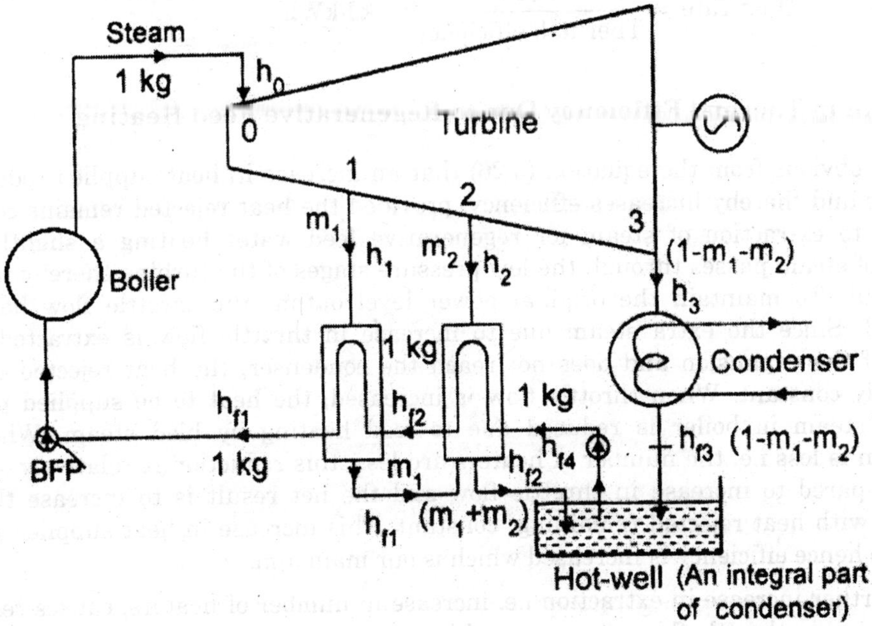

Fig. 3.16. With Plain Surface Heaters, Drain Fed to Hot-well.

At Heater No. 1 : $m_1 h_1 + h_{f2} = h_{f1} + m_1 h_{f1}$

or
$$m_1 = \frac{h_{f1} - h_{f2}}{h_1 - h_{f1}} \quad (3.24)$$

At Heater No. 2 : $m_1 h_{f1} + m_2 h_2 + h_{f4} = h_{f2} + (m_1 + m_2) h_{f2}$

or
$$m_2 = \frac{m_1 h_{f2} + h_{f2} - h_{f4} - m_1 h_{f1}}{h_2 - h_{f3}} \quad (3.25)$$

At Hot-well : $h_{f4} = (m_1 + m_2) h_{f2} + (1 - m_1 - m_2) h_{f3}$

Putting the value of h_{f4} we can calculate m_2.

The expression for themral efficiency will be on the above line.

3.9 OPTIMUM FEED WATER TEMPERATURE AND SAVING IN HEAT RATE

There is an optimum feed water temperature by which only the regenerative feed heating can be carried out very efficiently. The criterion governing the number of heaters is one of economic justification. To understand the optimum feed water temperature let us first define heat rate. *Heat rate is defined as the heat supplied per unit of work obtained, expressed generally as kJ/kWh. Sometimes, it is also termed as heat consumption.*

$$\text{Heat rate} = \frac{\text{Heat supplied} \times 3600}{\text{Heat supplied} - \text{Heat rejected}} \text{ kJ/kWh} \quad (3.26)$$

or
$$\text{Heat rate} = \frac{3600}{\text{Thermal efficiency}}; \text{ kJ/kWh} \quad (3.27)$$

3.9.1 Gain in Thermal Efficiency Due to Regenerative Feed Heating

It is obvious from the equation (3.26) that an increase in heat supplied reduces the heat rate and thereby increases efficiency, provided the heat rejected remains constant. But due to extraction of steam for regenerative feed water heating a slightly lower amount of steam passes through the low pressure stages of the turbine thereby reducing the output. To maintain the original power level output the throttle flow has to be increased. Since the extra steam due to increase in throttle flow is extracted in the middle of the expansion and does not reach the condenser, the heat rejected remains practically constant. When throttle flow is increased, the heat to be supplied per unit mass of steam in boiler is reduced due to feed heating by bled steam. When this extraction is less i.e. the number of heaters are less, this reduction is relatively small as heat compared to increase in throttle flow and the net result is to increase the heat supplied with heat rejected remaining constant. This increase in heat supplied reduces heat rate hence efficiency is increased which is our main aim.

A further increase in extraction i.e. increase in number of heaters, causes relatively less increase in throttle flow, as steam at higher pressure stages is extracted, which has higher enthalpy and therefore, is required in smaller quantities. The net effect is that the heat supplied in boiler per unit mass of steam reduces at a faster rate than the rate of increase in throttle flow. Hence the heat supplied is less which increases the heat rate thereby reducing efficiency. This imposes a limit on the extent to which regenerative feed heating is beneficial. The abve result is based on thermodynamical point of view only.

Reduction in heat rate obtained with various numbers of stages of extractions is shown in Fig. 3.17. These curves show that the improvement in efficiency due to stage extraction increases as the final feed water temperature increases upt a certain point, after that it decreases. It is obvious, that as that as the number of stages of extraction increases, the most efficient final feed water temperature increases. Also as the number of heaters increases the reduction in heat rate increased i.e. the gain in thermal efficiency increases.

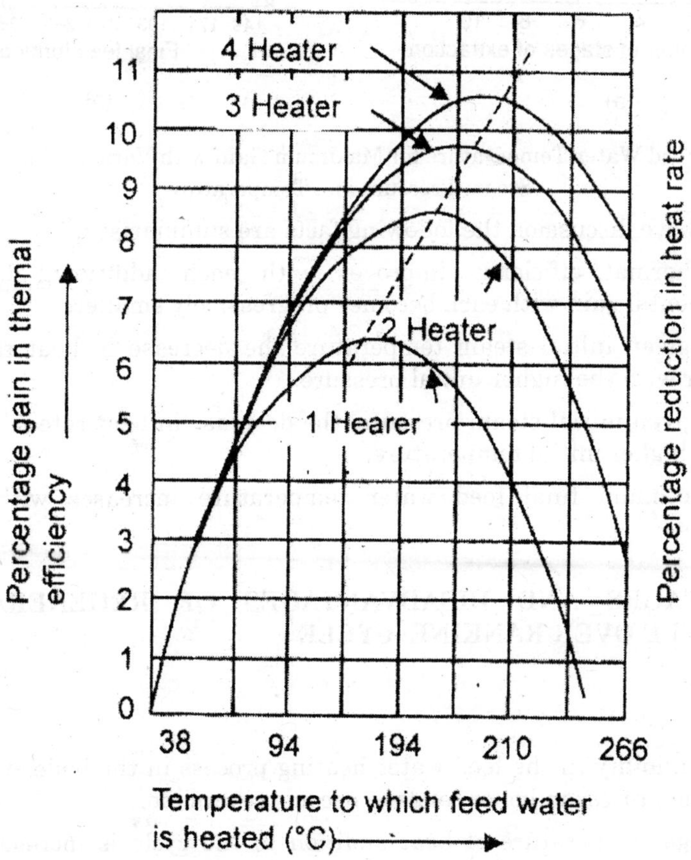

Fig. 3.17. Gain in Thermal Efficiency due to Regenerative Feed Heating.

Fig. 3.18, shows the effect of throttle pressure on temperature to which the feed water should be heated to obtain the maximum gain with stage extraction. It is seen from these curves that the most efficient final feed water temperature increases quite rapidly as heaters are added from the first four heaters while any further extraction above four affect the efficient final feed temperature only slightly. The higher the boiler pressure or the lower the condenser pressure, the better will be the gain affected by the regenerative cycle.

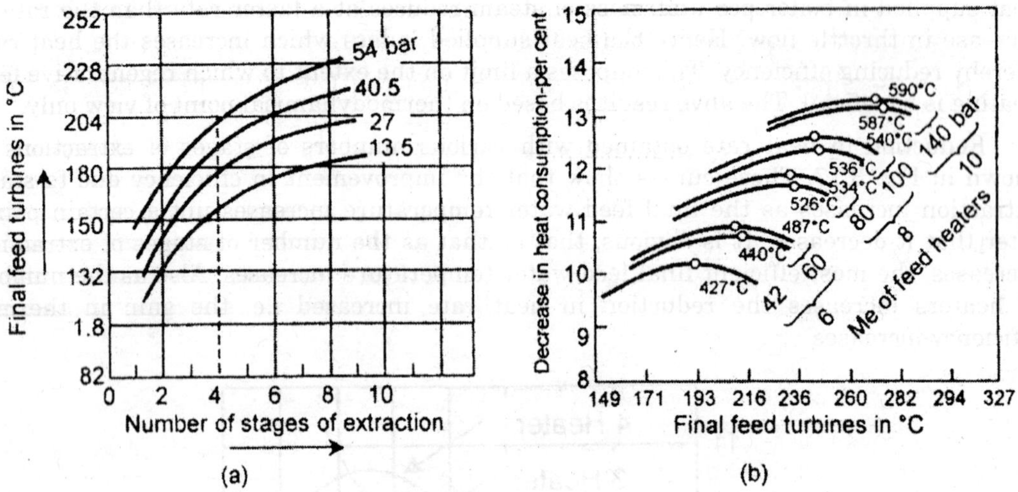

Fig. 3.18. Final Feed Water Temperature for Maximum Gain with Various Number of Stages and Steam Pressure and Temperature.

From the above discussion the following facts are summarised.

(i) The thermal efficiency improves with each additional heater but the incremental gain with each becomes progressively smaller.

(ii) For a given initial steam temperature the decrease in heat rate (kJ/kWh) is greater with the higher initial pressure.

(iii) For a given initial steam pressure, the decrease in heat rate is slightly smaller with a higher initial temperature.

(iv) The optimum final feed water temperature increases with initial steam pressure.

3.10 ADVANTAGES AND DISADVANTAGES OF REGENERATIVE FEED HEATING CYCLE OVER RANKINE CYCLE

Advantages

(i) Irreversibility in the feed water heating process in the boiler is minimised i.e. heating process in boiler tends to become reversible.

(ii) Average temperature of heat addition to the cycle is increased due to high temperature of feed water entering the boiler, thus increasing the thermal efficiency. As the numbers of heaters increases, the thermal efficiency increases as shown in Fig. 3.17 and 3.18.

(iii) Temperature ranges in the boiler are reduced, thus reducing thermal stresses set-up in the boiler.

(iv) The efficiency with which the extracted steam rejects heat to feed water is appreciably high, near about 98 percent. So the extraction of steam is advantageous because most of the extraction is at low pressure stages of the turbine where the steam expands with relatively low efficiency. The extracted quantity is limited by the feed water temperature.

(v) Less amount of steam is passed through the low pressure stages, so blade height will be less resulting in reduction in the cost of L.P. turbine.

(vi) Improved turbine drainage due to many extractions reduces erosion due to moisture.

(vii) It requires a reduced condenser capacity (i.e. condenser size) i.e. less costly condenser.

(viii) Heat rate is reduced.

Disadvantages

(i) It increases the complications of the plant. The addition of heaters, require greater maintenance.

(ii) The heaters and pipings are costly. High pressures heaters are much more expensive than the low pressure heaters. The gain in thermal efficiency is not must compared to the heavier costs.

(iii) It requires a larger boiler capacity for a given power.

Note: In the absence of precise information regarding the actual temperatures of the feed water entering and leaving the heaters and of the condensate temperatures, the following assumptions should always be made in solving problems:

(i) The bled steam just condenses, i.e. gives up its superheat (if any) and all its latent heat only. The condensed steam therefore leaves the heater at the saturation temperature corresponding to the bleeding pressure.

(ii) The feed water is heated to saturation temperature at the pressure of the bled steam.

And either:

(iii) (a) The condensed bled steam is returned to the system by separate pumps at a point down steam from the heater i.e. drain is fed to the high pressure side or (b) The condensed bled steam discharges into the hot-well, thereby arising the hot-well temperature; or (c) The condensed bled steam is cascaded down and discharges finally into the hot-well with or without drain cooler or condenser.

3.11 FEED WATER HEATER

A feedwater heater is basically a heat exchanger where heat is transmitted from the steam to the feedwater either by mixing the two fluid streams (open feedwater heaters) or without mixing them (closed feed water heaters).

3.11.1. Open feed water heaters

An open (or direct contact) feed water heater is basically a mixing chamber, where the steam extracted from the tubine mixes with the feed water exiting the pump. Ideally, the mixture leaves the heater as a saturated liquid at the heater pressure.

Fig. 3.19 Show a open feed water heater. Fig. 3.10 shows the schematic diagram and the associated T-s diagram for a regenerative vapour power cycle having one open feedwater heater.

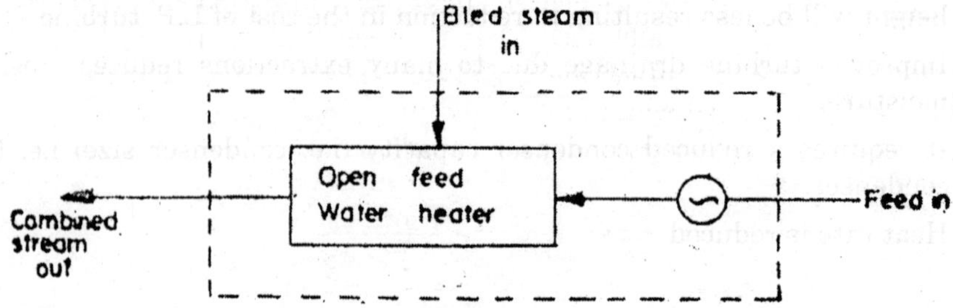

Fig. 3.19 Open feed water heater

3.11.2. Closed feed water heaters

Another type of feed water heater frequently used in steam power plants is the closed feed water heater, in which heat is transferred from the extracted steam to the feed water without any mixing taking place. The two streams now can be at different pressures, since they do not mix. The diagrams of Fig. 3.20 show two different schemes for removing the condensate from closed feed water heater. In Fig. 3.20(a) this is accomplished by means of a pump whose function is to pump the condensate forward to a higher pressure point in the cycle. In Fig. 3.20(b), the condensate is allowed to pass through a trap into a feed water heater operating at a lower pressure or into the condenser. A trap is a type of valve that permits only liquid to pass through a region of lower pressure.

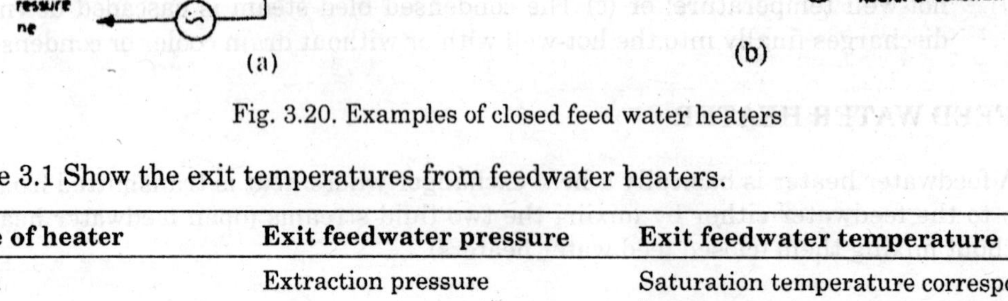

Fig. 3.20. Examples of closed feed water heaters

Table 3.1 Show the exit temperatures from feedwater heaters.

Type of heater	Exit feedwater pressure	Exit feedwater temperature
Open	Extraction pressure	Saturation temperature corresponding to extraction pressure (saturated liquid)
Closed	Boiler pressure	Saturation temperature corresponding to extraction pressure (compressed liquid)

Fig. 3.21 shows diagrammatically the temperature rise of the feed water in the various types of feed heaters. The thermal performance of feed heater may be accessed by the difference in temperature between the water leaving the heater and the saturation temperature corresponding to the heater steam pressure; this difference is termed as *terminal temperature difference* (T.T.D.) Direct contact heaters result in zero TTD. For a plain condensing (indirect contact) heater, T.T.D. is about 6°C. It is desirable to have T.T.D. equal to zero (i.e. feed water is heated to saturation temperature corresponding to the pressure of steam) or even small negative value. This is possible only with the use of desuperheater when the heater is supplied with highly superheated steam. In desuperheater, the temperature of the steam is reduced to near saturation temperature before exhausting it to the main heater for condensation. This is done because of the fact that the heat transfer coefficient from the superheated steam to the feed water is about ten times less than that of saturated steam. The desuperheater may be a separate heat exchanger for reducing the steam to near saturation temperature before exhausting it to the main heater for condensation. Alternatively, it may integral with the main heater and consist simply of baffling aroung the tubes carrying the feed water to the outlet of the heater. Separate drain coolers (i.e. it cools condensed steam and thus heats up feed water) are in common use but these are also incorporated with h.p. heaters. *The inclusion of desuperheating and drain cooling sections within the shell complicates feed heater design and construction but this design reduces capital cost and so it is in current use.* Again, to access the thermal performance of a drain cooler the tool used is the difference in temperature between the feed water entering and drain leaving (T.T.D.). For drain cooler, T.T.D. is equal to about 8°C. The flash type drain cooler is one in which drains entering it flash into steam and are then condensed. The integral drain cooler is water to water type heat exchanger.

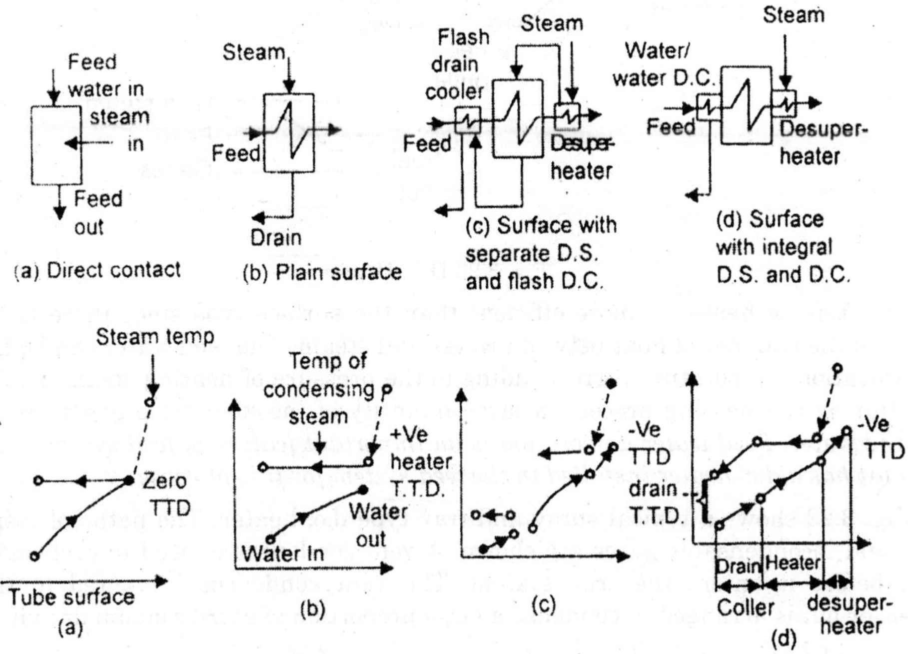

Fig. 3.21. Temperature Rise in Feed Heaters.

3.12 DIRECT CONTACT (D.C.) HEATERS

In this type of heater, there is a direct mixing of bled steam and feed water; the steam gives up its latent heat and thus condenses and as a result the feed water gets heated. In order to help this mixing process to take place in as short a time and in as small a volume as possible, the feed water entering at the top of the heater is atomised by spray nozzles and the resulting droplets fall through the steam atmosphere and cascade over perforated plates or tray stages (Fig. 3.22).

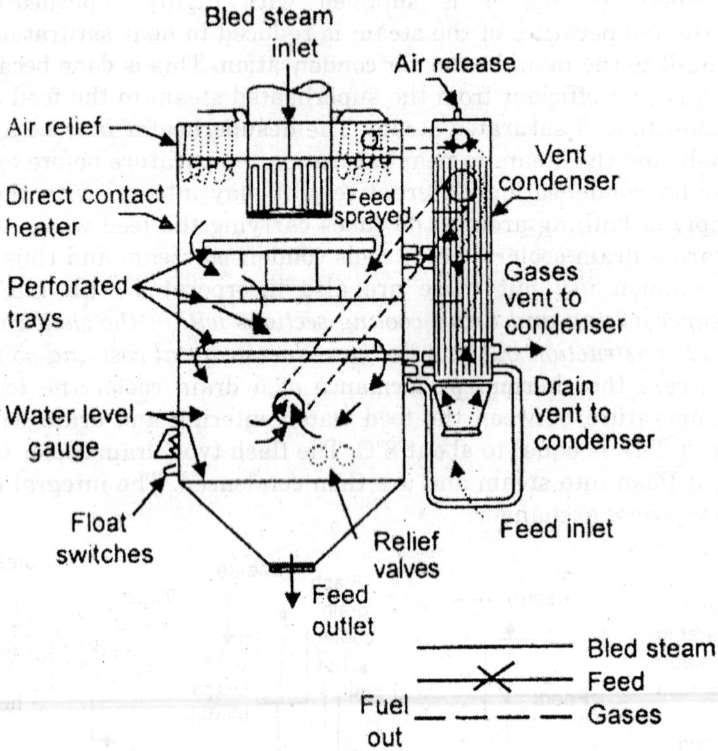

Fig. 3.22 D.C. Heater

This type of heater is more efficient than the surface type since there is no metal barrier to the transfer of heat between water and steam. The feed water can be heated to the saturation temperature corresponding to the pressure of heating steam, i.e. T.T.D. is zero. During the heating process, a large quantity of incondensable gases are released from the water. *Feed water de-aeration is an important feature of feed system design and each unit has a d.c. heater installed in the feed system for this purpose.*

Fig. 3.22 shows a typical spray and tray type d.c. heater. The paths of steam, feed water and incondensable gases are shown. A vent condenser is used to preheat the feed water before it enters the tray system. The vent condenser is vented to the main condenser but is arranged to condense a large proportion of carrier steam drawn into it.

A novel feature of d.c. heater scheme is the use of glandless pumps with water fillea squirrel cage motors and the arrangement of heaters are one above the other in the train at a sufficient elevation to ensure gravity flow between them at all loads saving pumping work. Due to use of glandless pump, there is no maintenance of gland and no risk of air leakage at the glands. There are the following other benefits from using d.c. heater instead of surface heaters:

(i) The elimination of non-ferrous materials removes the possibility of copper pickup in the feed system.

(ii) Better deaeration is possible as each d.c. heater acts to some an extent as a deaerator.

The turbine is protected against flooding of a d.c. heater by an isolating valve in the feed line before the first heater, closed automatically by a heater high water level.

Particularly, during periods of changing load, the risk of flooding may be further reduced by designing the system to operate at full load with no stored water in the heaters. The use of water seal prevents the steam flow between heaters under all operating conditions.

3.13 SURFACE HEATERS

In the surface type of feed heater, the feed water passes through *hairpin tubes* in two, four or occassionally six flows (three hairpin tubes in series), while the steam is passed over the tubes. Baffles are provided to direct the flow of steam through the tube nest. Baffles also act as support plates and allow thermal expansion of tubes.

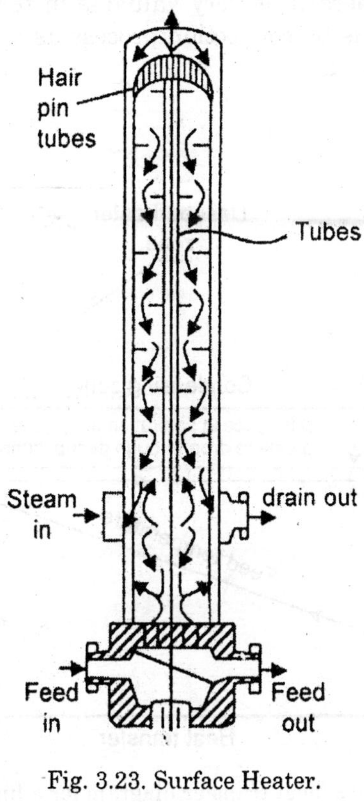

Fig. 3.23. Surface Heater.

The condensate is drained from the bottom of the heater shell to be cascaded to the next lower heater train, or to be pumped forward ahead of the heater into the feed line towards high pressure side. Pumping forward (high pressure side) mixes the drains with the feed water at the outlet temperature of the heater and is, therefore a more efficient method of drains heat utilisation. The disadvantage lies in requiring a pump for each heater, its maintenance and the risk of air leakage at the glands.

Fig. 3.23 shows a h.p. heater of surface type. The current practice on large high pressure systems is to install desuperheating and drain cooling section within the shell of the heater. Although the inclusion of desuperheating and drain cooling sections within the shell complicates the heater design and construction, but it reduces the overall capital cost. In such a heater, there are three zones namely *desuperheating, condensing and drain cooling.*

3.13.1 Desuperheating Zone

This zone is the last feed heater zone through which the superheated bled steam enters and the feed water passes before leaving the heater. Fig. 3.24 shows temperature/heat transferred diagram for a typical high pressure heater. It is obvious from the diagram that when the feed water enters the zone it has been heated in the condensing zone to within a few degrees of the saturation temperature, corresponding to the bled steam pressure at the entry to the feed heater. Although, there is an addition of only a small percentage of the total heat transferred in the heater by desuperheating zone, this small rise in feed water temperatre is very valuable in terms of thermal economy. The heat transferred in this zone is by convection process as the superheated steam may be treated to behave as a normal gas.

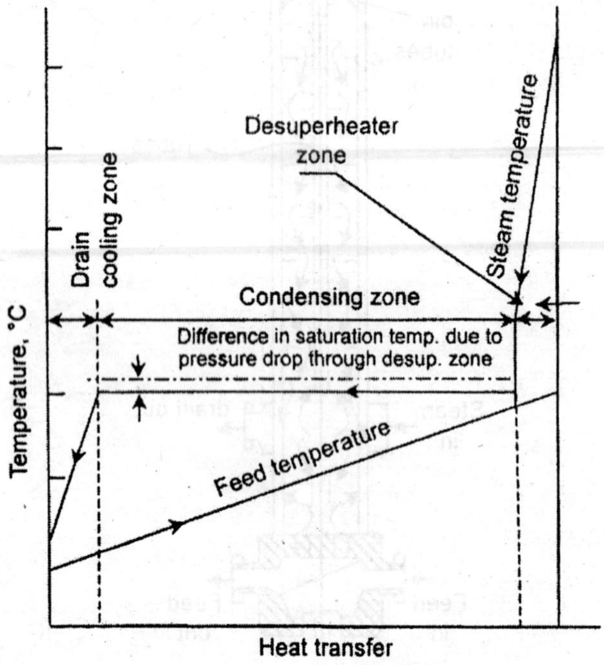

Fig. 3.24. Temperature /heat transfer diagram for a high pressure heater.

The normal practice is to allow the steam to leave the desuperheating zone with some residual superheat of about 27.8°C. The temperature gradent between the bulk steam and the feed water temperature is shown in the top of the figure. Since the steam leaves with some residual superheat, so the tube wall temperature is above saturation temperature of the bled steam and no condensation takes place on the tube wall under normal conditions. That is why there is no problem of formation of droplets of condensed steam and being carried into the condensing zone by the relatively high velocity steam and thus causing impingement attack.

3.13.2 Condensing Zone

The major heating of feed water takes place in the condensing zone as the overall heat transfer coefficient in this zone is very high. The main design requirement of large heaters is to provide good steam distribution to the condensing surface with minimum pressure loss. Under normal load running conditions the ingress of non-condensable gases is unlikely, however, there is a likelihood of its accumulation after shut-down during two shift operation. Under this condition, continuous venting is necessary to help the steam distribution and to clear any non-condensable gases accumulated.

The use of diaphragm plates serve a dual purpose, in addition to supporting the tube nest, they also prevent the accumulation of condensate from running down the outsides of the tubes and forming a thick film which impairs the heat transfer coeffcient.

3.13.3 Drain Cooling Zone

The drain cooling zone provides a fairly high heat transfer rate provided the drain are not reheated by heat transferred through the shrouds around the drain cooler convection one. The reheating of the drain is possible if there is no desuperheating section. In general, the desuperheating zone is larger than the drain cooling zone and it is possible to have the heater shell flooded with condensate to above the drain cooler inlet. By this way, a stagnant pool of condensate is formed around the drain cooler shrouds causing them to follow the temperature gradient in the drain cooler. As a result of this the problem of drains reheat is eliminated provided the shell drain valves are kept closed.

The present practice is to keep the high pressure heater up side down. The reduces the cost of feed pipe work and has operational advantage that it permits drainage of the tubes during prolong shut down periods and thus reducing feed water contamination and metal pick-up.

3.13.4 Automatic Heater Bypassing

There is a risk of flooding of high pressure surface heater owing to bursting of tubes, leaking tube/tube-plate joints or drain blockage. Precautions must be taken to prevent water from a flooded heater shell being passed back into the turbine through the bled steam pipes. In order to circumvent this problem, an automatic by-pass system around the high pressure heater or group of heaters is installed, so that, when a dangerously high water level occurs in a heater shell, the unit is automatically isolated on the feed side from the feed pump (Fig. 3.25). By this way, any further ingress of water into the heater shell is prevented.

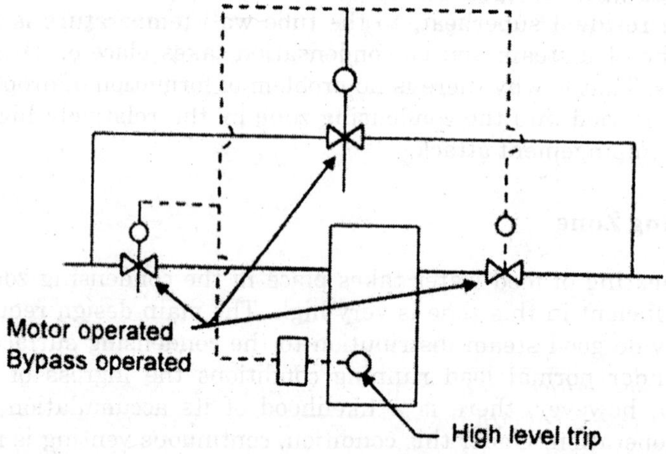

Fig. 3.25. Feed Heater Bypass System

3.13.5 Tube Material

The tube material used for the heating surface in L.P. heaters is usually non-ferrous and copper and brass are used extensively because of their satisfactory heat transfer. The 90/10 grade of cupro-nickel alloy is in service in some h-p heaters with feed water temperature up to 224°C but steel is used for high temperature.

3.14 DEAERATORS

There are corrosive effects of water which contains oxygen dissolved from the atmosphere and thus produces detrimental effects on boiler tubes, steam pipes and other items of the plant. Therefore, one requirement of a modern feed water system is that is must supply water, which has an almost zero oxygen content, to the feed pumps for feeding the boiler. In general, for boilers operating at a pressure in excess of 70 bar, the feed water should have an oxygen content of less than 0.007 mg/kg which is possible by physical de-aeration but it is possible to reduce this value to 0.007 mg/kg by chemical dosing. *Since a feed system is closed after the boiler feed pump, all physical deaeration must be carried out before the feed pump suction.*

The amount of oxygen that can be retained in solution is proportional to the partial gas pressure and depends inversely on the temperature of the water. It is to be noted that condensers act as deaerators and reduce the oxygen content in condensate to a very low level, for the use of independent de-aerator is unnecessary for many feed systems. However, in order to insure that the amount of dissolved oxygen in a system is at a minimum, it is usual to install a de-aerator to supplement the condenser. The solubility of air in water decreases with a rise in temperature and a reduction in pressure. Therefore, a deaerator is based on the principle that the amount of gas in solution in water is proportional to the pressure of that gas in the water and on the surface of the water. Theoretically, if water is boiling, no gas can exist in the water on its surface and any gases in the water are driven out. It has been found in practice that heating the water to boiling point de-aerated the water only if sufficient time is allowed to elapse. However, the water can be de-aerated if it is boiled rapidly and voilently; in order to achieve this the water is heated under reduced pressure (vacuum condition). If the steam used to remove the gases in this way is taken from the intermediate stage of the turbine, the heat required to remove the gases will also help in increase the thermal efficiency of the system by imparting heat to the feed water.

Another important requirement of feed water system is to have a reservoir or storage tank of suitably de-aerated water having sufficient capacity, about one-sixth of the MCR feed flow, situated above the feed pump suction to meet any sudden or instantaneous demands for feed water by the boiler. It is convenient to carry out the de-aeration of the feed water in a de-aerating heater which acts as a direct-contact feed heating stage in the system, and to combine the de-aerating heater with the storage tank. In other words, *a de-aerator (a direct contact heater) which removes oxygen content from the feed water using bled steam, combined with the storage tank and arranged to an elevated position on the suction of feed pump ensures a positive suction head under all conditions of operation to feed de-aerated water to the boiler.*

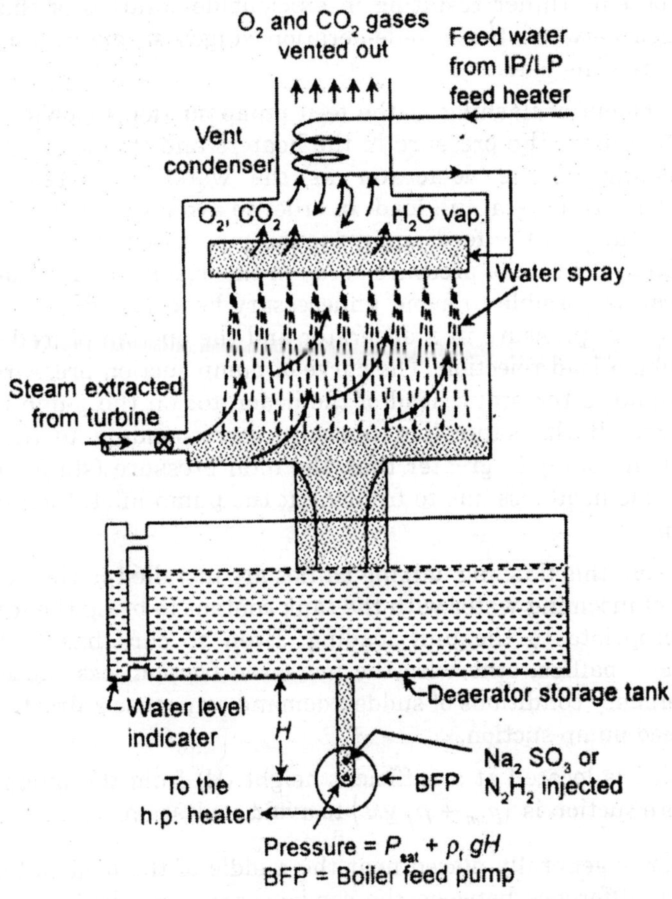

Fig. 3.26. De-aerator and Storage Tank.

Fig. 3.26 shows a typical arrangement of a thermal de-aerator and storage tank. The de-aeratoion and heating is carried out in the two domes fitted to the storage tank. Bled steam is allowed to enter the storage tank above the water level from the top and flows along the steam space in the storage tank and up into the de-aerating domes. Condensate passes through the surface type devaporizer and enters and de-aerating dome at the top. There are essentially two stages in which the de-aeration and heating are carried out in the de-aerating dome. There are a spray stage and a tray stage. As the condensate enters, it passes through many small sprayers in parallel into the upper steam portion of the dome. The condensate are dispersed by the sprays into thin conical films which eventually coalesce to form small droplets which pass through the steam atmosphere into the trays below. The trays are arranged in series with the sprays on the condensate side and perform a similar duty so that a thin film of water is cascaded over the weir at the inner and outer edges and droplets cascade between trays through the perforations in them.

Approximately, 95 per cent of the heating the de-aeration takes place in the spray stage to full load. But with the decrease of load, the flow and pressure loss across the spray portion decreases resulting loss in efficiency in the atomization of the water and a reduced rate of heat transfer. With the decrease in water flow cascading over the trays the water films become thiner resulting in efficient de-aeration of that stage. The net result is a balance between the two-de-aeration stages so giving an almost constant efficiency with decreasing load.

There is a problem of flashing at the feed pump suction following a load rejection. During a rapid load drop, the pressure in the heaters and the de-aerator tends to drop. This causes flashing in the de-aerator as the water is stored at boiling point corresponding to the pressure at full load. In order to avoid the flashing in the pump, the rate at which the water in the feed pump suction pipe gains static head as it descends must be greater than the rate of pressure decay in the deaerator. The suction pipe should be as near vertical as possible to avoid unnecessary head loss due to friction. Fig. 3.27 shows the variation in pressure in a deaerator and the suction of feed pump positioned beneath it following a load rejection. The curve of pump suction pressure is greater, by a constant figure equal to the static head of the deaerator on the pump than the curve of de-aerator pressure. If the saturation pressure corresponding to the temperature of water arriving at the pump is greater than the total pressure (static head + de-aerator pressure of that time-head loss due to friction) at the pump inlet, then flashing will take place in the pump.

In order to avoid this flashing, mixing baffles are installed in the tank to promote an adequate mixing of incoming water with the tank content to bring the total quickly to the temperature appropriate to the unit loading. Even so care has to be taken in the installation of these baffles, otherwise the increased friction loss caused by heavy flow rates during emergency conditions of sudden demand can cause a drastic reduction in the pressure at the feed pump suction.

The deaeration is located at a sufficient height (H) from the pump location so that the pressure before suction is $\left(p_{sat} + \rho_f\, gH\right)$ to avoid cavitation.

The deaerator is generally placed near the middle of the feed water system so that the total pressure difference between the condenser and the boiler is shared equitably between the condensate pump and boiler feed pump. The feed water heaters before the deaerator are often termed as high pressure (HP) heaters and those after deaerator and termed as low pressure (LP) heaters.

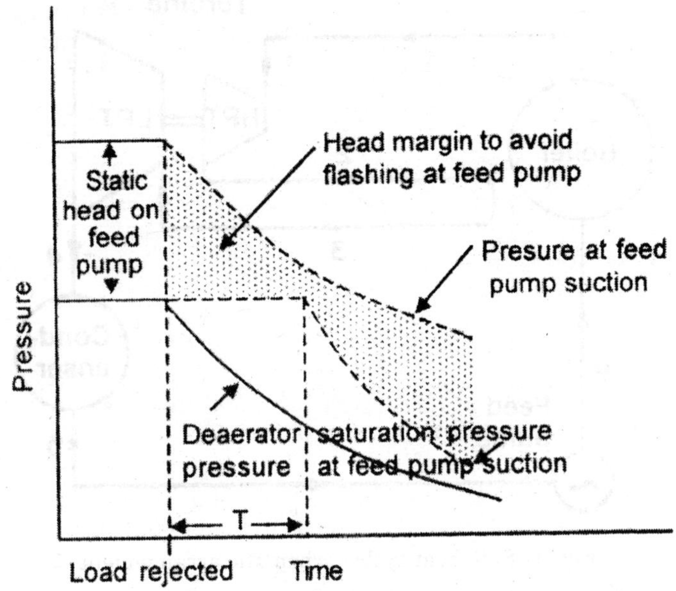

Fig. 3.27. Pressure Variation air Deaerator.

3.15 REHEAT CYCLE

In the last section we noted that the thermal efficiency of a Rankine cycle can be improved by increasing the inlet pressure, decreasing the exhaust pressure, and superheating the supply steam. For a given maximum superheating temperature, which is usually determined by the strength characteristics of construction materials, increasing the steam generator pressure causes a decrease in the quality of the steam leaving the prime mover. This is undesirable if the prime mover is a turbine because moisture content of more than 10 to 12 percent causes serious erosion of turbine blades. The reheat cycle has been developed to take advantage of the increased efficiency with higher pressures, and yet avoid excessive moisture in the low pressure stages of the turbine.

The basis of this cycle is an attempt to approach Carnot cycle efficiency by adding heat in increments at the highest possible temperature level. The steam is permitted to expand part of the way in the turbine and is then returned to the boiler, where it is heated again (reheated) and subsequently re-expanded through the turbine. Although there is no theoretical limit to the number of stages of reheat that can be employed in a cycle, two or at most three are used in practice.

The reheat cycle is shown schematically in Fig. 3.28 and on a T-s and h-s diagram in Fig. 3.29.

In this cycle, the steam is first expanded to some intermediate pressure in the turbine and is then reheated at constant pressure in the reheater, after which it is expanded to the exhaust pressure. The turbine may be considered as having two stages higher pressure and low pressure.

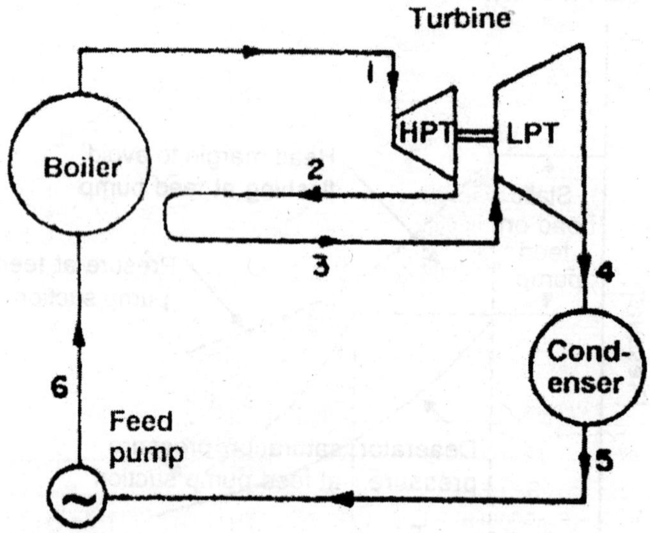

Fig. 3.28. Reheat cycle - schematic arrangement

Heat is added at three places; in the boiler (process in the boiler (process 6→8); in the super heater (process 8 →1) and in the reheater (process 2→3). Work produced at two places: In the high pressure turbine (1→2) and in the low pressure turbine (3→4), point e is the exhaust state without reheating. Heat is rejected in the condenser, and work is supplied to the pump. Thus, for this cycle, we have

$$\text{Heat supplied} = (h_1 - h_6) + (h_3 - h_2) \tag{3.28}$$

$$\text{Work done} = (h_1 - h_2) + (h_3 - h_4) \tag{3.29}$$

$$\text{Heat rejected} = (h_5 - h_4) \tag{3.30}$$

$$\text{Pump work} = (h_6 - h_5) \approx -v_{f3}(p_1 - p_3) \tag{3.31}$$

$$\text{Thermal} \quad \eta = \frac{\text{Turbine work} - \text{Pump work}}{\text{Heat supplied}}$$

$$= \frac{(h_1 - h_2) + (h_3 - h_4) - (h_5 - h_4)}{(h_1 - h_6) + (h_3 - h_2)} \tag{3.32}$$

In practice, reheat equipment is economically justifiable only in high capacity plants (say above 100 bar). Due to reheating there is very little gain in efficiency because the average temperature at which heat is supplied is not greatly changed? The chief advantage is in decreasing the moisture content in the low pressure stages of the turbine to a safe value (Fig. 3.29). If metals could be found that would enable one to superheat the steam, using Rankine cycle would be more efficient than the reheat cycle, and there would be no need for reheat cycle.

The main disadvantage of reheating is that for very high pressure cycle, doubling reheating is used. In other words the benefits or reheat are gained at the expense of the reheat equipment i.e. cost of the plant increases. Condenser capacity due to increased dryness fraction is also increased.

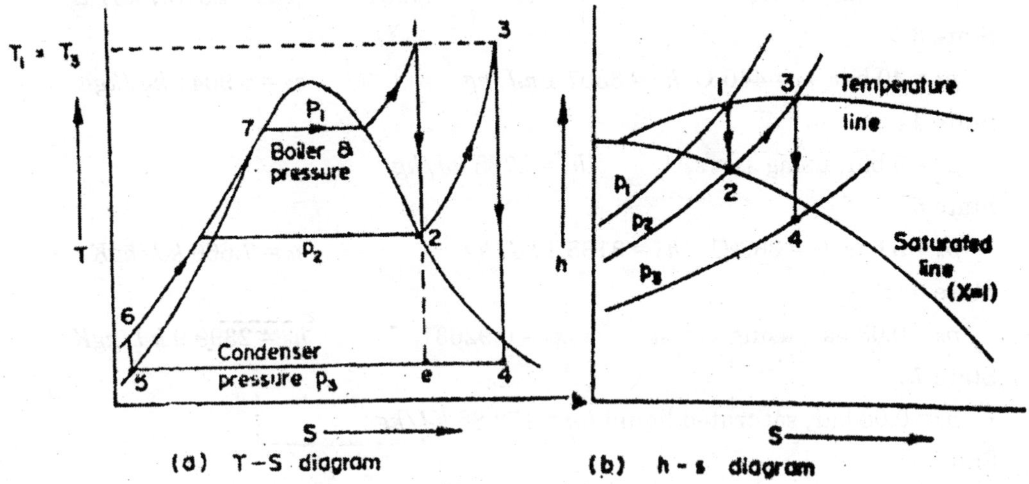

Fig. 3.29. T-s and h-s diagrams.

ILLUSTRATIVE EXAMPLES

3.1 Steam at 320 bar, 520°C leaves the steam generator of an ideal Rankine cycle modified to include three turbine stages with reheat between the stages. The reheat pressures are 40 bar and 5 bar respectively, and steam enters the second stage turbine at 440°C and third stage turbine at 360°. The condenser pressure is 0.08 bar. Determine for the cycle (a) the net work per unit mass of steam flowing, in *kJ/kg* (b) the thermal efficiency.

Solution

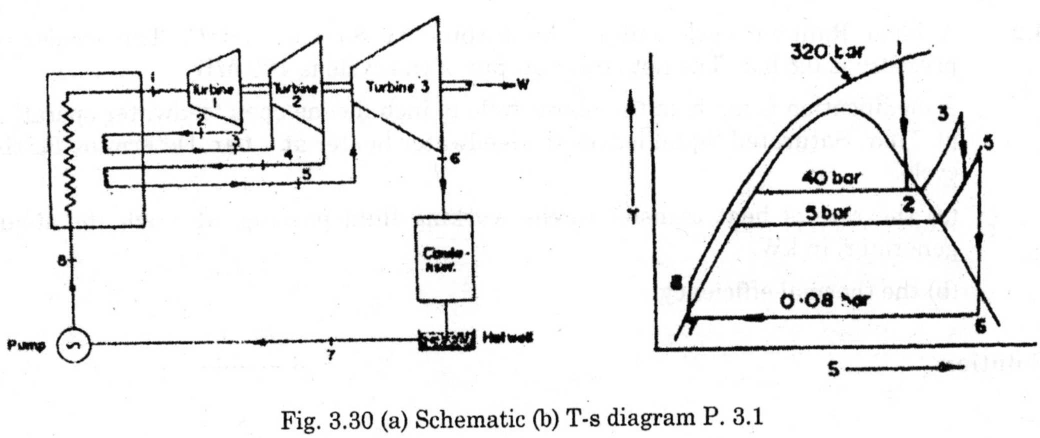

Fig. 3.30 (a) Schematic (b) T-s diagram P. 3.1

State 1 :

$p_1 = 320$ bar, $t_1 = 520°C h_1 = 3133.7\ kJ/kg$ $s_1 = 5.8357\ kJ/kgK$

State 2 :

$p_2 = 40$ bar, using $s_1 = s_2$ $x_2 = 0.9284$ $\therefore h_2 = 2678.7\ kJ/kg$

State 3 :

$p_3 = 40$ bar, $t_3 = 440°C$, $h_3 = 3307.1\ kJ/kg$ $s_3 = 6.9041\ kJ/kgK$

State 4 :

$p_4 = 5$ bar, using $s_3 = s_4$ $h_4 = 2785\ kJ/kg$

State 5 :

$p_5 = 5$ bar, $t_5 = 360°C$. $h_5 = 3188.4\ kJ/kg$ $s_5 = 7.666\ kJ/kgK$

State 6 :

$p_6 = 0.08$ bar, using $s_5 = s_6$ $x_6 = 0.92631$ $h_6 = 2399.9\ kJ/kgK$

State 7 :

$p_7 = 0.08$ bar, saturated liquid $h_{f7} = 173.88\ kJ/kg$

State 8 :

$$h_8 = h_7 + v_{f7}(p_8 - p_7)$$

$$= 173.77 + 1.0084 \times 10^{-3}(320 - 0.08) = 206.14\ kJ/kg$$

(a) $$W_T = (h_1 - h_2) + (h_3 - h_4) + (h_5 - h_6)$$

$$= (3133.7 - 2678.7) + (3307.1 - 2785) + (3188.4 - 2399) = 1765.6\ kJ/kg$$

$$W_p = h_8 - h_7 = 32.26\ kJ/kg$$

$$W_{cycle} = W_T - W_p = 1765.6 - 32.26 = 1733.3\ kJ/kg$$

(b) $$Q = (h_1 - h_8) + (h_3 - h_2) + (h_5 - h_4) = 3959.4$$

$$\eta = \frac{W}{Q} = \frac{1733.3}{3959.4} = 0.438 = 43.8\%$$ **Ans.**

3.2 A ideal Rankine cycle enters the turbine at 80 bar, 480°C. The condenser pressure is 0.8 bar. The net power output of the cycle is 100 *MW*.

A modification is made in the above cycle to include one open feedwater operating at 7 bar. Saturated liquid leaves the feedwater heater at 7 bar. Determine for the cycle.

(a) the rate of heat transfer to the working fluid passing through the steam generator, in kW.

(b) the thermal efficiency.

Solution

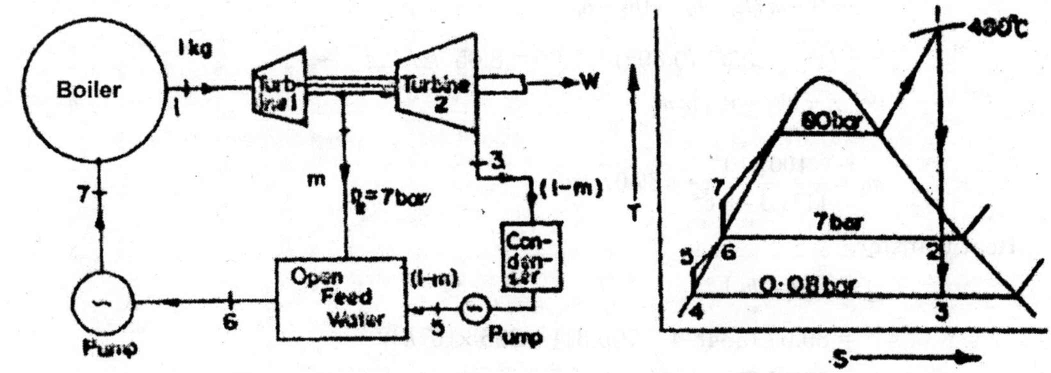

Fig. 3.31 (a) Schematic diagram (b) T-s diagram P. 3.2

State 1 :

$p_1 = 80$ bar, $t_1 = 480°C$ $h_1 = 3348.4$ kJ/kg $s_1 = 6.6586$ kJ/kgK

State 4 :

$p_3 = 0.08$ bar $= p_4$ $h_{f4} = 173.88$ kJ/kg $v_{f4} = 1.008 \times 10^{-3}$ m^3/kg

State 2 :

$p_2 = 7$ bar

Using $s_1 = s_2'$ $x_2 = 0.9895$, $h_2 = 2741.8$ kJ/kg

State 3 :

Using $s_1 = s_2 = s_3$ $h_3 = 2082.9$ kJ/kg

State 5 :

$$h_3 = h_{f4} + v_{f4}(p_5 - p_4)$$

$$= 173.88 + 1.008 \times 10^{-3} (7 - 0.08) \times 100 = 174.58 \; kJ/kg$$

State 6 :

$p_6 = 7$ bar, $h_{f6} = 697.22$ kJ/kg $v_{f6} = 1.108 \times 10^{-3}$ m^3/kg

State7:

$$h_7 = h_{f6} + v_{f6}(p_7 - p_6)$$

$$= 697.22 + 1.108 \times 10^{-3} (80 - 7) \times 100 = 705.31 \; kJ/kg$$

Applying mass and energy rate balances to the control volume enclosing the open feed water heater, the fraction m of flow extracted at location 2 is

$$m = \frac{h_{f6} - h_5}{h_2 - h_5} = \frac{697.22 - 174.58}{2741.8 - 174.58} = 0.2036 \; kg$$

(a) $$W_T = (h_1 - h_2) + (1 - m)(h_2 - h_3)$$

$$= (3348.4 - 2741.8) + (1 - 0.2036) \times (2741.8 - 2082.9)$$

$$= 1131.3 \; kJ/kg$$

$$W_p = (1-m)(h_5 - h_4) + (h_7 - h_6)$$

$$= (1-0.2036)(0.698) + 8.09 = 8.65 \; kJ/kg$$

$$W_{cycle} = (W_T - W_p) \times m_1$$

$$m_1 = \frac{100 \times 10^3}{1131.3 - 8.65} = 89.07 \; kg/s$$

Heat transfer,

$$Q = m_1(h_1 - h_7)$$

$$= 89.07\,(3348.4 - 705.31) = 235 \times 10^3 \, kW$$

$$= 235 \; MW \quad \text{Ans.}$$

(b) $$\eta = \frac{W}{Q} = \frac{100 \times 10^3}{235 \times 10^3} = 0.425 = 42.5\%$$ Ans.

3.3 A steam power plant operates between a boiler pressure of 42 bar and a condenser pressure of 0.035 bar. If the Rankine cycle of the above limits is modified to include one feed heater, calculate the cycle efficiency and specific steam consumption.

Solution

At 42 bar, $t_1 = 235.2°C$ and at 0.035 bar, $t_3 = 26.7°C$

Therefore, $$t_2 = \frac{253.2 + 26.7}{2} = 140°C$$

Selecting the nearest saturation pressure from the tables gives the bleed pressure as 3.5 bar i.e. $t_2 = 138.9°C$.

Therefore, $$mh_2 + (1-m)h_4 = 1 \times h_6$$

$$mh_2 + h_4 - mh_4 = h_6$$

$$m(h_2 - h_4) = h_6 - h_4$$

$$m = \frac{h_6 - h_4}{h_2 - h_4} = \frac{h_{f6} - h_{f4}}{h_2 - h_{f4}} = \frac{h_{f6} - h_{f3}}{h_2 - h_{f3}}$$

At 3.5 bar, $h_{f6} = 584 \; kJ/kg$

At 0.035 bar, h_{f3} or $h_{f4} = 112 \; kJ/kg$

and $s_1 = s_2 = s_{3'}$

the dryness fraction

is obtained as $x_2 = 0.829$ and $x_3 = 0.696$

Therefore, $$h_2 = h_{f2} + x_2 h_{fg2}$$

$$= 584 + 0.829 \times 2148 = 2364 \; kJ/kg$$

$$h_3 = h_{f3} + x_3 h_{fg3} = 112 + 0.696 \times 2438 = 1808 \ kJ/kg$$

Therefore, $$m = \frac{h_6 - h_4}{h_2 - h_4} = \frac{h_{f6} - h_{f3}}{h_2 - h_{f3}}$$

$$= \frac{584 - 112}{2364 - 112} = 0.21 \ kg$$

Neglecting the second feed pump term i.e. $h_6 = h_7$. We have

Heat supplied in boiler = $h_1 - h_{f6}$

$$= 2800 - 584 = 2216 \ kJ/kg$$

Total work output $W = W_{12} + W_{23}$

$$= (h_1 - h_2) + (1 - m)(h_2 - h_3)$$

$$= (2800 - 2364) + (1 - 0.21)(2364 - 1808) = 876 \ kJ/kg$$

$$\eta = \frac{\text{Work done}}{\text{Heat supplied}} = \frac{876}{2216} = 0.396 = 39.6\%$$

Specific steam consumption $= \frac{3600}{W} = \frac{3600}{876} = 4.12 \ kg/kWh$ Ans.

3.4 A power plant operates on a regenerative vapour power cycle with one closed feedwater heater. Steam enters the first turbine stage at 120 bar, 520°C and expands to 10 bar, where some of the steam is extracted and diverted to a closed feed water heater. Condensate leaving the feedwater herater as saturated liquid at 10 bar passes through a trap into the condenser. The feedwater leaves the heater at 120 bar with a temperature of 170°C. The condenser pressure is 0.06 bar. Determine for the cycle (a) the thermal efficiency and (b) the net power developed in kW for a mass flow rate of 10^6 Kg/h into the first stage turbine.

Solution

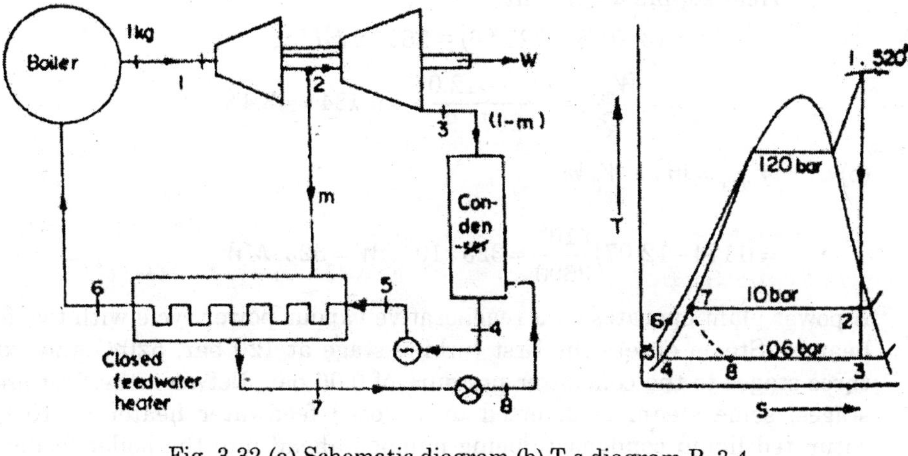

Fig. 3.32 (a) Schematic diagram (b) T-s diagram P. 3.4

State 1 :

$p_1 = 12$ bar, $t_1 = 520^oC, h_1 = 3401.8\ kJ/kg$ $\quad s_1 = 6.55\ kJ/kgK$

State 2 :

$p_2 = 10$ bar using $s_1 = s_2$ $\quad x_2 = 0.9931$ $\quad h_2 = 2764.2\ kJ/kg$

State 3 :

$p_3 = 0.06$ bar, *using* $s_1 = s_3$ $\quad x_3 = 0.7727$ $\quad h_3 = 2018.3 kJ/kg$

State 4 :

$p_4 = 0.06$ bar, $h_{f4} = 151.53\ kJ/kg$

State 5 :

$$h_5 = h_4 + v_{f4}(p_5 - p_4)$$

$$=151.53 + 1.0064 \times 10^{-3} (120 - 0.06) \times 100 = 151.53 + 12.07 = 163.6\ kJ/kg$$

State 6 :

$p_6 = 120$ bar, $t_6 = 170^oC$ using interpolation in table, $h_6 = 725.86\ kJ/kg$

State 7 :

$p_7 = 10$ bar, $h_{f7} = 762.81\ kJ/kg$

State 8 : Throttling process $h_{f7} = h_{f8} = 762.81\ kJ/kg$

(a) The mass m is calculated with the help of mass and energy balance in the closed feed water heater,

$$\therefore \quad m = \frac{h_6 - h_{f5}}{h_2 - h_{f7}} = \frac{725.86 - 163.6}{2764.2 - 762.81}$$

$$W_T = (h_1 - h_2) + (1 - m)(h_2 - h_3)$$

$$= (3401.8 - 2764.2) + (1 - 0.2809) \times (2764.2 - 2018.3) = 1174\ kJ/kg$$

$$W_p = h_3 - h_{f4} = 163.6 - 151.53 = 12.07\ kJ/kg$$

Heat supplied = $h_1 - h_6$

$= (3401.8 - 725.86) = 2675.9\ kJ/kg$

$$\eta = \frac{W_{net}}{Q} = \frac{1174 - 12.07}{2675.9} = 0.434 = 43.4\%$$

(b) $\quad W_{cycle} = (W_T - W_p)m$

$$= (1174 - 12.07)\frac{\times 10^6}{3600} = 323 \times 10^3\ kW = 323\ MW$$ Ans.

3.5 A power plant operates on a regenerative vapour power cycle with two feedwater heaters. Steam enters the first turbine stage at 120 bar, 520°C and expands in three stages to the condenser pressure of 0.06 bar. Between the first and second stages, some steam is diverted to a closed feedwater heater at 10 bar, with saturated liquid condensate being pumped ahead into the boiler feedwater line. The feedwater leaves the closed heater at 120 bar, 170°C. Steam is extracted between the second and third turbine stages at 1.5 bar and fed into an open feed water heater operating at that pressure. Saturated liquid at 1.5 bar leaves the open feed water heater. Determine for the cycle (a) the thermal efficiency and (b) the net power developed in *kW* for a mass flow rate of 10^6 *kgh* into the first stage turbine.

Solution :

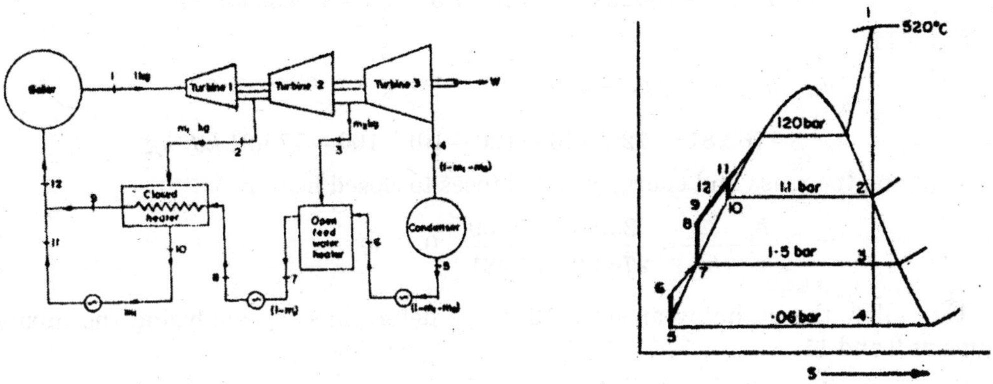

Fig. 3.33 (a) Schematic (b) T-s diagram P. 3.5

State 1:

$p_1 = 120$ bar, $t_1 = 520°C$

$h_1 = 3401.8\ kJ/kg$ $s_1 = 6.5555\ kJ/kgK$

State 2 :

$p_2 = 10$ bar, Using $s_1 = s_2$

$x_2 = 0.9931$ $h_2 = 2764.2\ kJ/kg$

State 3 :

$p_3 = 1.5$ bar, Using $s_1 = s_3$, $x_3 = 0.8847$ $h_3 = 2436.9\ kJ/kg$

State 4 :

$p_4 = 0.06$ bar, Using $s_1 = s_4$, $x_4 = 0.7727$ $h_4 = 2018.3\ kJ/kg$

State 5 :

$p_5 = 0.06\ bar$, $h_{f5} = 151.53\ kJ/kg$

State 7 :

$p_7 = 1.5$ bar, $h_{f7} = 467.11\ kJ/kg$

State 10 :

$p_{10} = 10$ bar, $h_{f10} = 762.81\ kJ/kg$

State 9 :

$p_9 = 120$ bar, $t_9 = 170°C$ $h_9 = 725.86\ kJ/kg$

State 6 :

$$h_6 = h_{f5} + v_{f5}(p_6 - p_5)$$

$$=151.33 + 1.0064 \times 10^{-3}\,(1.5 - 0.06) \times 100 = 151.67\ kJ/kg$$

State 8 :

$$h_8 = h_{f7} + v_{f7}(p_8 - p_9)$$

$$= 467.11 + 1.0528 \times 10^{-3}\,(120 - 1.5) \times 100 = 479.59\ kJ/kg$$

State 11:

$$h_{11} = h_{f10} + v_{f10}(p_{11} - p_{10})$$

$$= 762.81 + 1.1273 \times 10^{-3}\,(120 - 10) \times 100 = 775.21\ kJ/kg$$

Applying the mass and energy rate balances to closed heater

$$m_1 = \frac{h_9 - h_8}{h_2 - h_{10}} = \frac{725.86 - 479.59}{2764.2 - 762.81} = 0.1230$$

This allow the enthalpy at state 12, to be determined by analyzing the mixing of steams at 9 and 11.

$$(1 - m_1)h_9 + m_1 h_{11} - h_{12} = 0$$

$$\therefore \quad h_{12} = (0.8770)\,(725.86) + 0.1230 \times 775.21 = 731.93\ kJ/kg$$

Next energy and mass rate balances for the open heater

$$m_2 h_3 + (1 - m_1 - m_2)h_6 - (1 - m_1)h_7 = 0$$

$$m_1 = \frac{(1 - m_1)(h_7 - h_6)}{h_3 - h_6}$$

$$= \frac{0.8770(467.11 - 151.67)}{2436.9 - 151.67} = 0.1211$$

$$W_T = (h_1 - h_2) + (1 - m_1)(h_2 - h_3) + (1 - m_1 - m_2)(h_3 - h_4)$$

$$= (3401.8 - 2764.2) + 0.8770(2764.2 - 2436.9) + 0.7559$$

$$(2436.9 - 2018.30)$$

$$= 1241.1\ kJ/kg$$

$$W_T = (h_1 - h_2) + (1 - m_1)(h_2 - h_3) + (1 - m_1 - m_2)(h_3 - h_4)$$

$$= (3401.8 - 2764.2) + 0.8770(2764.2 - 2436.9) + 0.7559$$

$$(2436.9 - 2018.30)$$

$$= 1241.1\ kJ/kg$$

$$Q = h_1 - h_{12} = 3401.8 - 731.93 = 2669.9\ kJ/kg$$

$$\eta = \frac{W_T - W_p}{Q} = \frac{1241.1 - 12.58}{2669.9} = 0.46 = 46\%$$

Ans.

$$W_{cycle} = \dot{m}_1 \left[W_T - W_p\right]$$

$$= \frac{10^6}{3600}[1241.1 - 12.58] = 341 \ MW \qquad \text{Ans.}$$

3.6 A power plant operates on a regenerative vapour power cycle with two feed water heaters, a closed one and an open one. Steam enters the first turbine stage at 80 bar, 480°C, and expands to 20 bar. Some steam is extracted at 20 bar and fed to the closed feedwater heater. The remainder expands through the second stage turbine to 3 bar, where an additional amount is extracted and fed into the open feedwater heater, which operates at 3 bar, the steam expanding through the third stage turbine leaves at the condenser pressure of 0.08 bar. Feedwater leaves the closed heater at 205°C, 80 bar, and condensate leaving as saturated liquid at 20 bar is trapped into the open heater. Saturated liquid at 3 bar leaves the open feedwater heater. The net power output of the cycle is 100 MW. Determine for the cycle (i) The thermal efficiency. (ii) The mass flow rate of steam entering the first turbine; in *kg/h*

(b) Modify the regenerative vapour power cycle to include reheat at 20 bar. The portion of the flow that passes through the second turbine stage is reheated to 440°C before it enters the second turbine stage. Determine the thermal efficiency of the modified cycle.

Solution :

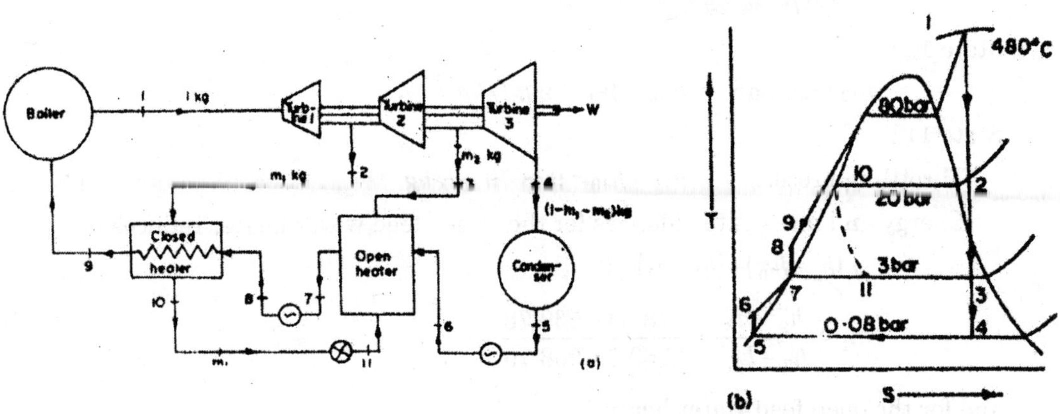

Fig. 3.34 (a) Schematic (b) T-s diagrams P. 3.6

State 1 :

$p_1 = 80$ *bar*, $t_1 = 480°C$, $h_1 = 3348.4 \ kJ/kg$ $s_1 = 6.6586 \ kJ/kgK$

State 2 :

$p_2 = 20$ bar, Using $s_1 = s_2$ $h_2 = 2963.5 \ kJ/kg$

State 3 :

$p_3 = 3$ bar, Using $s_1 = s_3$, $x_3 = 0.9373$, $h_3 = 2589.6 \ kJ/kg$

State 4 :

$p_4 = 0.08$ bar, $s_1 = s_4$, $x_4 = 0.7944$, $h_4 = 2082.9\ kJ/kg$

State 5 :

$p_5 = 0.08$ bar, $h_{f5} = 173.88\ kJ/kg$

State 7 :

$p_7 = 3$ bar, $h_{f7} = 561.47\ kJ/kg$

State 6 :

$$h_6 = h_{f5} + v_{f5}(p_6 - p_5)$$

$$= 173.88 + 1.0084\times10^{-3}\,(3 - 0.08)\times 100 = 173.88 + 0.294 = 174.17\ kJ/kg$$

State 8 :

$$h_8 = h_{f7} + v_{f7}(p_8 - p_7)$$

$$= 561.47 + 1.0732\times10^{-3}\,(80 - 3)\times100 = 561.47 + 8.26 = 569.73\ kJ/kg$$

State 9 :

$p_9 = 80$ bar, $t_9 = 205°C$

With the help of interpolation or using the formula

$h_9 = 878.34\ kJ/kg$

State 10 :

$p_{10} = 20$ bar, $h_{f10} = 908.79\ kJ/kg$

State 11 :

Throttling process, $h_{11} = h_{10} = 908.79\ kJ/kg$

Energy and mass rate balances for the closed feed water heater reduces to

$$m_1(h_2 - h_{10}) + (h_8 - h_9) = 0$$

$$m_1 = \frac{h_9 - h_8}{h_2 - h_{10}} = \frac{878.34 - 569.73}{2963.5 - 908.79}$$

and for the open feed water heater

$$m_2h_3 + (1 - m_1 - m_2)h_6 + m_1h_{11} - h_7 = 0$$

$$m_2 = \frac{m_1(h_6 - h_{11}) + (h_f - h_6)}{h_3 - h_6}$$

$$= \frac{0.1502(174.17 - 908.79) + (561.47 - 17.47)}{2589.6 - 174.17} = 0.1147$$

$$W_T = (h_1 - h_2) + (1 - m_1)(h_2 - h_3) + (1 - m_1 - m_2)(h_3 - h_4)$$

$$= (3348.4 - 2963.5) + 0.8498(2963.5 - 2589.6) + 0.7351(2589.6 - 2082.9)$$

$$=1075.1\ kJ/kg$$

$$W_p = (h_8 - h_7) + (1 - m_1 - m_2)(h_6 - h_5)$$

$$= (569.73 - 561.47) + (0.7351)\ (174.17 - 173.88) = 8.47\ kJ/kg$$

$$Q = h_1 - h_9 = 3348.4 - 878.34 = 2470.1\ kJ/kg$$

$$\eta = \frac{W_T - W_p}{Q} = \frac{1075.1 - 8.47}{2470.1} = 0.4318 = 43.18\%$$ Ans.

(ii) $W_{cycle} = \dot{m}_1 \left[W_T - W_p \right]$

$$\dot{m}_1 = \frac{100 \times 10^3}{1066.63} = 3.375 \times 10^5\ kg/h$$

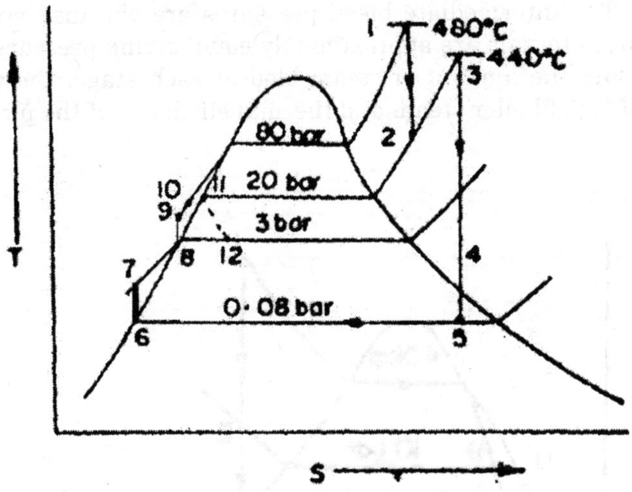

Fig. 3.35 T-s diagram P. 3.6

(b) State 3 : $p_3 = 20$ bar, $t_3 = 440$°C

$h_3 = 3335.5\ kJ/kg$ $s_3 = 7.254\ kJ/kgK$

State 4 : $p_4 = 3$ bar, Using $s_3 = s_5$ $h_4 = 2839.5\ kJ/kg$

State 5 : $p_5 = 0.08$ bar, $s_3 = s_5$

$x_5 = 0.8723$ $h_5 = 2270.1\ kJ/kg$

$$m_2 = \frac{m_1 (h_7 - h_{12}) + (h_8 - h_7)}{h_4 - h_7}$$

$$= \frac{0.1502(174.17 - 908.79) + (561.47 - 174.47)}{2839.5 - 174.17} = 0.1029$$

$$W_T = (h_1 - h_2) + (1 - m_1)(h_3 - h_4) + (1 - m_1 - m_2)(h_4 - h_5)$$

$$= (3348.4 - 2963.5) + 0.8498\ (3335.5 - 2839.5) + (0.7459)\ (2839.5 - 2270.1)$$

$$= 1231.1\ kJ/kg$$

$$W_p = (h_9 - h_8) + (1 - m_1 - m_2)(h_7 - h_6)$$

$$= (569.73 - 561.47) + (0.7459)(174.17 - 173.88) = 8.48 \ kJ/kg$$

The total heat added during superheating and reheating is expessed per unit mass entering the turbine as

$$Q = (h_1 - h_{10}) + (1 - m_1)(h_3 - h_2)$$

$$= (3348.4 - 878.34) + (0.8498)(335.5 - 2963.5) = 2786.2 \ kJ/kg$$

$$\eta = \frac{W_T - W_p}{Q} = \frac{1231.1 - 8.48}{2786.2} = \frac{1222.62}{2786.2} = 0.4388 = 43.88\% \qquad \text{Ans.}$$

3.7 In a regenerative steam cycle employing two closed feedwater heaters the steam is supplied to the turbine at 40 bar and 550°C is exhausted to the condenser at 0.035 bar. The intermediate bleed pressures are obtained such that saturation temperature intervals are approximately equal giving pressures of 10 bar and 1.1 bar. Calculate the amount of steam bled at each stage, the work output of the plant in *kJ/kg* of boiler steam and thermal efficiency of the plant.

Solution:

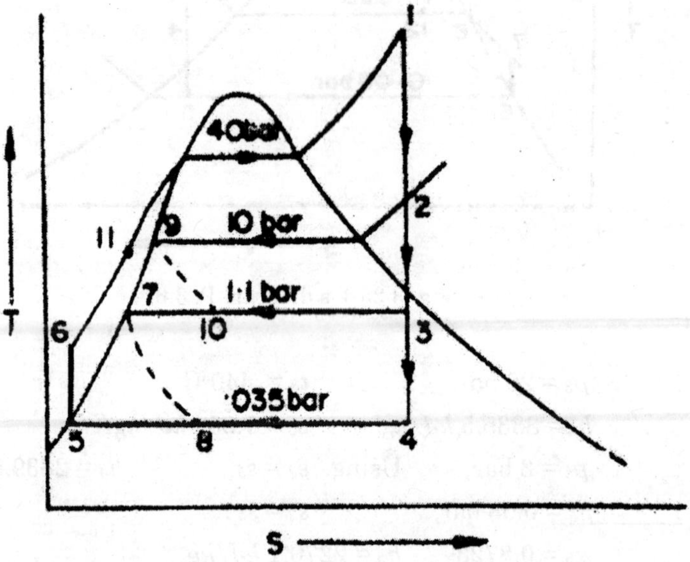

Fig. 3.36 T-s diagram P. 3.7

State 1 :

$h_1 = 3445 \ kJ/kg$ $\qquad s_1 = 7.089 \ kJ/kgK$

State 2 :

$s_1 = s_2$ $\qquad h_2 = 3032.9 \ kJ/kg$

State 3 :

$s_1 = s_2 = s_3$ $\qquad x_3 = 0.961$ $\qquad h_3 = 2591 \ kJ/kg$

State 5 :

$$h_{f5} = 112\ kJ/kg$$

State 4 :

$s_1 = s_2 = s_3 = s_4$ $\quad x_2 = 0.824$ $\quad h_2 = 2117\ kJ/kg$

For the throttling process

$$h_9 = h_{10} = 763\ kJ/kg\ = h_{11}$$

For the throttling process

$$= h_7 = h_8 = 429\ kJ/kg = h_6$$

m_2 is calculated by balancing mass energy in heater

$$m_1 = \frac{h_{11} - h_6}{h_2 - h_9} = \frac{763 - 429}{3032.9 - 763} = 0.147$$

Similarly for second heater

$$m_2 = 0.124$$

Work output $= (h_1 - h_2) + (1 - m_2)(h_2 - h_3) + (1 - m_1 - m_2)(h_3 - h_4)$

$= (3445 - 3032.9) + 0.853\ (3032.9 - 2591) + 0.729\ (2591 - 2117)$

$= 1134.5\ kJ/kg$

$$Q = h_1 - h_{11} = 3445 - 763 = 2682\ kJ/kg$$

$$\eta = \frac{W}{Q} = \frac{1134.5}{2682} = 0.43 = 42.3\%$$ Ans.

3.8 Steam enters the first turbine stage of a vapour power cycle with reheat and regeneration at 320 bar, 600°C, and expands to 80 bar, A portion of the flow is diverted to a closed feedwater heater at 80 bar, and the remainder is reheated to 500°C before entering the second turbine stage. Expansion through the second turbine stage occurs to 10 bar, where another portion of the flow is diverted to a second closed feedwater heater at 10 bar. The remainder of the flow expands through the third turbine stage to 1.5 bar, where a portion of the flow is diverted to an open feedwater heater operating at 1.5 bar, and rest expands through the fourth turbine stage to the condenser pressure of 0.06 bar. Condenser leaves each closed feedwater heater as saturated liquid at the respective extraction pressure. The condensate streams from the closed heaters each pass through traps into the next lower pressure feedwater heater. Saturated liquid leaving the open heater is pumped to the steam generator pressure. For isentropic operation of the turbine stages and pumps

(a) Sketch the layout of the cycle and number of principal state points.

(b) determine the thermal efficiency of the cycle.

(c) Calculate the mass flow rate into the first turbine stage, in *kg/h*, for a net power output of 500 *MW*.

Solution

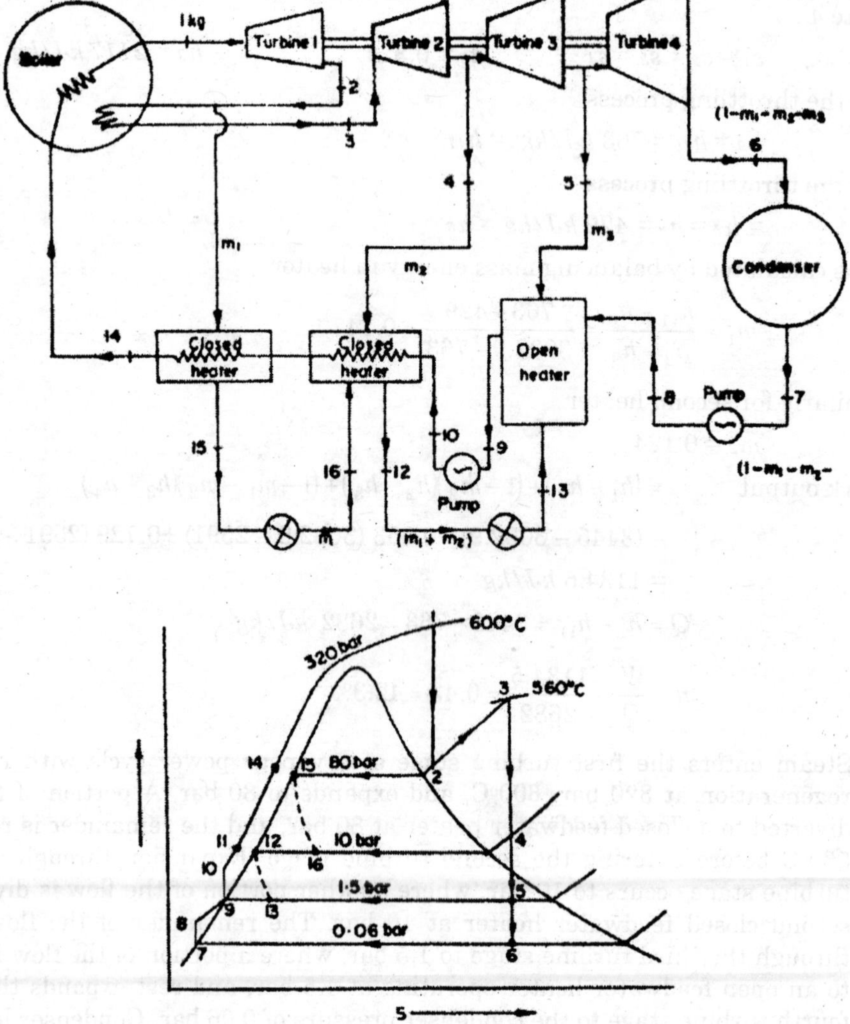

Fig. 3.37 (a) Schematic diagram (b) T-s diagram P.3.8

State 1 :

$p_1 = 320$ bar, $t_1 = 600°C$, $h_1 = 3424.6\ kJ/kg$ $s_1 = 6.1858\ kJ/kgK$

State 2 :

$p_2 = 80$ bar, $s_1 = s_2$ $h_2 = 3022.3\ kJ/kg$

State 3 :

$p_3 = 80$ bar, $t_3 = 560°C$ $h_3 = 3545.3\ kJ/kg$ $s_3 = 6.9072\ kJ/kgK$

State 4 :

$p_4 = 10$ bar, Using $s_3 = s_4$ $h_4 = 2934\ kJ/kg$

State 5 :

$p_5 = 1.5$ bar, Using $s_3 = s_5$ $x_5 = 0.9464,$ $h_5 = 2572\ kJ/kg$

State 6 :

$p_6 = 0.06$ bar, $s_6 = s_3$ $x_6 = 0.81776,$ $h_6 = 2127.2\ kJ/kg$

State 7 :

$p_7 = 0.06$ bar, $h_{f7} = 151.53\ kJ/kg$

State 8 :

$$h_8 = h_{f7} + v_{f7}(p_8 - p_7) = 151.53 + 1.0064\times10^3(1.5 - 0.06)\times100$$

$$= 151.67\ kJ/kg$$

State 9 :

$p_9 = 1.5$bar, $h_{fg} = 467.11\ kJ/kg$

State 10 :

$$h_{10} = h_{fg} + v_{fg}(p_{10} - p_9)$$

$$= 467.11 + (1.0528\times10^3\ (320 - 1.5)\times 100) = 500.64\ \underline{kJ/kg}$$

State11:

$p_{11} = 320$bar, $t_{11} = t_{sat}$ [At 10 bar $t_{sat} = 180°C$] $h_{11} = 781.7\ kJ/kg$

State12 :

$p_{12} = 10$ bar, $h_{f12} = 762.81\ kJ/kg$

State13 :

Throttling process $h_{12} = h_{13} = 762.81$

State14 :

$p_{14} = 320$ bar, $t_{14} = t_{sat}$ [At 80 bar, $t = 295°C$] $h_{14} = 1303.8\ kJ/kg$

State 15 :

$p_{15} = 80$ bar, $h_{f15} = 1316.6\ kJ/kg$

State16:

$h_{16} = h_{15} = 1316.6\ kJ/kg$

Mass and energy balances 80 bar closed heater

$$m_1 = \frac{h_{14} - h_{11}}{h_2 - h_{15}} = \frac{1303.8 - 781.7}{3022.3 - 1316.6} = 0.3061$$

Smlary for the 10 bar closed heater

$$m_2 = \frac{(h_{11} - h_{10}) + m_1(h_{12} - h_{16})}{h_4 - h_{12}}$$

$$= \frac{(781.7 - 500.64) + (0.306061)(762.81 - 1316.60)}{2934 - 762.81} = 0.0514$$

For the open heater

$$m_3 = \frac{(h_9 - h_8) + (m_1 - m_2)(h_8 - h_{13})}{h_5 - h_8}$$

$$= \frac{(467.11 - 151.67) + (0.3575)(151.67 - 762.81)}{2572 - 151.67} = 0.0401$$

$$W_T = (h_1 - h_2) + (1 - m_1)(h_3 - h_4) + (1 - m_1 - m_2)(h_4 - h_5) + (1 - m_1 - m_2 - m_3)(h_5 - h_6)$$

$$= (3424.6 - 3022.3) + 0.6939(3545.3 - 2934) + 0.6425 (2934 - 2572) + 0.6024(2572 - 2127.2 = 1327 \ kJ/kg)$$

$$W_p = (h_{10} - h_9) + (1 - m_1 - m_2 - m_3)(h_8 - h_7)$$

$$= (500.64 - 467.11) + 0.6024(151.67 - 151.53) = 33.61 \ kJ/kg$$

$$Q = (h_1 - h_{14}) + (1 - m_1)(h_3 - h_2)$$

$$= (3424.6 - 1303.8) + 0.6939 \ (3545.3 - 3022.3) = 2483.7 \ kJ/kg$$

$$\eta = \frac{W_T - W_p}{Q} = \frac{1327 - 33.61}{2483.7} = 0.521 = 52.1\%$$ Ans.

$$W_{cycle} = m_1 [W_T - W_p]$$

$$500 \times 1000 = m_1 [1327 - 33.61]$$

$$m_1 = 1.392 \times 10^6 \ kg/h$$ Ans.

EXERCISES

A. THEORY

3.1 Define the following terms: Work ratio, specific steam consumption, overall efficiency, isentropic efficiency and relative efficiency.

3.2 What is the work ratio and what is its physical significance?

3.3 What should be the properties of working fluid so that the Rankine cycle approaches the Carnot cycle.

3.4 Why is the Carnot cycle not a realistic model for steam power plants?

3.5 Why is excessive moisture in steam undersirable in steam turbines? What is the highest moisture content allowed?

3.6 How is moisture content related to quality?

3.7 Explain why the Rankine cycle rather than the Carnot cycle is used as a standard of reference for the performance of steam plants?

3.8 Compare the efficiency and specific steam consumption of ideal Carnot cycle and Rankine cycle as a function of boile pressure.

3.9 Discuss the effects of following parameters in a Rankine cycle.

(a) Steam temperature at inlet to the turbine; (b) Steam pressure at inlet to the turbine; (c) Steam pressure at exhaust.

3.10 How does the efficiency of the Rankine cycle compare to that of a Carnot cycle operating between the same temperature limits?

3.11 How do actual vapour power cycles differ from the idealized ones?

3.12 Prove by means of comparison of significant areas on the T-s chart that, for an ideal Rankine cycle with fixed initial and exhaust pressure, a cycle using superheated steam is more efficient than one using dry saturated steam.

3.13 What are the desirable properties of a fluid for use as working fluid in a heat engine plant?

3.14 Compare the pressures at the inlet and exit of the boiler for (a) actual and (b) ideal cycles.

3.15 During a regenerative process, some steam is extracted from the turbine and is used to heat the liquid water leaving pump. This does not seem like a smart thing to do since the extracted steam could produce some more work in turbine. How do you justify this action?

3.16 How do open feedwater heaters differs from closed feedwater heater?

3.17 Devise an ideal regenerative Rankine cycle that has the same thermal efficiency as the Carnot Cycle. Show the cycle on a T-s diagram.

3.18 Write advantages and disadvantages regenerative cycle over Rankine cycle.

3.19 Draw neat sketches of open feed water and closed feed water heaters.

3.20 Draw neat sketches of temperature rise diagrams of various feed heaters.

3.21 Draw neat sketches of DC heater and explain working.

3.22 Draw neat sketches of surface heater and explain working.

3.23 Draw temperature/Heat transfer distributor in surface heater.

3.24 Draw neat sketches of feed heater bypass system and explain workng.

3.25 Draw neat sketches of Deaerator and storage tank and explain working.

3.26 Draw pressure variation air deaerator.

3.27 Under what conditions does reheating increases the efficiency of the ideal Rankine cycle?

3.28 Can the efficiency of the ideal Rankine cycle decrease as a result of reheating? Explain.

B. UNSOLVED PROBLEMS

3.1 In a regenerative vapour power cycle with one open feed water heater. Steam enters the turbine at 80 bar, 480°C and expands to 7 bar, where some of the steam is extracted and directed to the open feedwater heater operating at 7 bar. The remaining steam expands through the second stage turbine to the condenser pressure of 0.008 MPa. Saturated liquid exits the open feedwater heater at 7 bar. The isentropic efficiency of each turbine stage is 85%. If the net power output of the cycle is 100 *MW*, determine (a) the thermal efficiency and (b) the mass flow rate of steam entering the first turbine stage, in *kg/h* [36.9%, 3.69x10^5 *kg/h*].

3.2 Consider an ideal steam regenerative cycle in which steam enters the turbine at 35 bar, 350°C and exhausts to the condenser at 10 *kPa* Steam is extracted from the turbine at 8 bar and also at 2 bar for heating the boiler feedwater in two open feedwater heaters. The feedwater leaves each heater at the temperature of the condensing steam. The appropriate pumps are used for the water leaving condenser and the two feedwater heaters. Calculate the cycle thermal efficiency and the net work per kg of steam. [36.9%, 878.6 *kJ/kg*].

3.3 Steam is the working fluid in an ideal Rankine cycle with superheat and reheat. Steam enters the first stage turbine at 8 *MPa*, 480°C, and expands to 0.7 *MPa*. It is then reheated to 440°C before entering the second stage turbine, where it expands to the condenser pressure of 0.008 *MPa*. The net power output is 100 *MW*. Determine (a) the thermal efficiency of the cycle (b) the mass flow rate of steam in *kg/h*, (c) the rate of heat transfer from the condenser in *MW*. [40.3%, 2.363 x 10^5 *kg/h*, 148 *MW*].

3.4 Consider a steam power plant that operates on a reheat Rankine cycle and has a net power output of 200 *MW*. Steam enters the high pressure turbine at 10 *MPa* and 500°C and the low pressure turbine at 1 *MPa* and 500°C. Steam leaves the condenser as a saturated liquid at a pressure of 10 *kPa*. The adiabatic efficiency of the turbine is 80 percent, and that of the pump is 95 percent. Show the cycle on *T-s* diagram and determine (a) the quality of the steam at the turbine exit, (b) the thermal efficiency of the cycle, and (c) The mass flow rate of the steam [187.5°C, 34.1%, 156.6 *kg*/s].

3.5 A steam power plant operates on an ideal Rankine cycle with two stages of reheat and has a net power output of 500 *MW*. Steam enters all three stages of the turbine at 500°C. The maximum pressure in the cycle is 15 *MPa*, and the minimum pressure 5 *kPa*. Steam is reheated at 5 *MPa* the first time and at 1 *MPa* the second time. Show the cycle on a T-s diagram and determine (a) the thermal efficiency of the cycle and (b) the mass flow rate of the steam. [45.5%, 268.5 *kg/s*].

3.6 A steam power plant operates on an ideal reheat regenerative Rankine cycle with one reheater and two open feedwater heaters. Steam enters the high pressure turbine at 10 *MPa* and 550°C and leaves the low pressure turbine at 5 *kPa*. Steam is extracted from the turbine 1 and 0.2 *MPa* and it is reheated to 500°C at a pressure of 0.8 *MPa*. Water leaves both feedwater heaters as a saturated liquid. Heat is transferred to the steam in the boiler at a rate of 10^6 *kJ/s*. Show the cycle on a T-s diagram and determine (a) the mass flow rate of steam through the boiler, (b) the net power output of the plant, and (c) the thermal efficiency of the cycle. [290.6 *kg/s*, 465.7 *MW*, 46.6%].

Chapter 4

GOVERNING OF STEAM TURBINES, COGENERATION, MIXED PRESSURE AND BINARY VAPOUR CYCLE

4.1 GOVERNING OF STEAM TURBINES

The quality of electric energy is determined by the voltage and frequency of alternating current, in particular, it is higher when these parameters are maintained constant with higher accuracy. Voltage control is effected mainly by excitation of the electric generator, which only deals with maintaining the specified consiant frequency.

Each turbine plant is provided with a speed governor which measures the rotational frequency of the turbine shaft and governs of system which correlates the external load of the generator and the torque on the turbine shaft.

The purpose of governing is to maintain the speed of the turbine constant regardless of the change in the load on the turbine. If the load on the turbine increase, the speed of the turbine will begin to decrease if the steam supply is not increased. On the other hand if the load on the turbine decreases, the speed of the turbine will begin to increase if the steam supply is not decreased. In the nutshell load is proportional to mass rate of flow. The purpose of the governing is to supply the steam to the turbine according to the load on the turbine and to maintain the speed of the turbine constant.

The following methods of governing the steam turbines are :

(1) Throttle governing.

(2) Nozzle control governing.

(3) By-pass governing.

(4) Combination of throttle and nozzle control governing or throttle and by pass governing.

4.1.1 Servomotor governing

With the development of steam turbine plants, the forces exerted by steam flow on governing members have increased so much that made the use of direct governing systems impossible. Intermediate mechanisms were introduced for actuating the heavily loaded steam governing members. A popular kind of such mechanisms is the piston type hydraulic drive, or servomotor. Its merits are a simple design, high reliability, ease of control, and the possibility of developing the full force in starting from rest.

A servomotor, or servodrive, is controlled by a cylindrical pilot valve which can be moved by smaller forces. In this type of governing system specified angular velocity is maintained without static error, i.e. the degree of irregularity of the system is zero. Such systems, i.e. those maintaining a constant rotational frequency, are called proportional plus integral governing systems.

A simple sketch of servomotor governing is shown in Fig. 4.1. and a transient processes occurring on a change from one steady state to another is shown in Fig. 4.2.

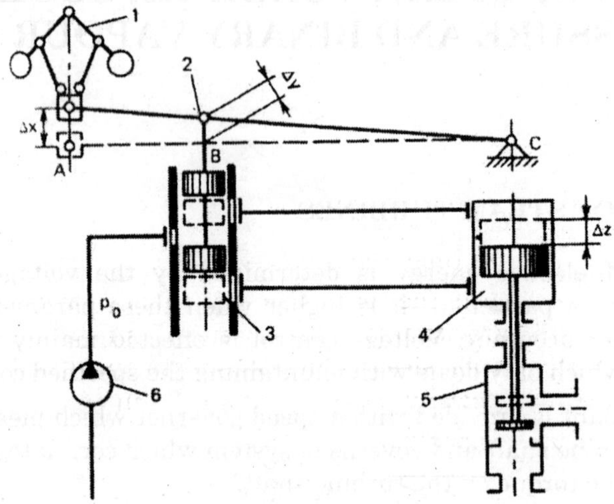

Fig. 4.1. Scheme of the governing system with servomotor.

Suppose that the external load has been shedded fully, which initiated an increase of angular velocity $\Delta \omega$ (Fig. 4.2). Accordingly, the governor sleeve moves upwards through a distance Δx and lifts the pilot valve by Δy. The pilot valve opens the inflow of liquid to the servomotor and the servomotor piston moves through a distance Δz and closes partially steam admission to the turbine. At a certain moment t_1, the admission valve reaches the idle-run position after which the angular velocity ceaces to increase. At that moment the pilot valve is still in a position above the mid position and the pressurized liquid continues to flow into the upper surface of the servomotor and the piston continues closing the valve until it is fully closed ($t = t_2$).

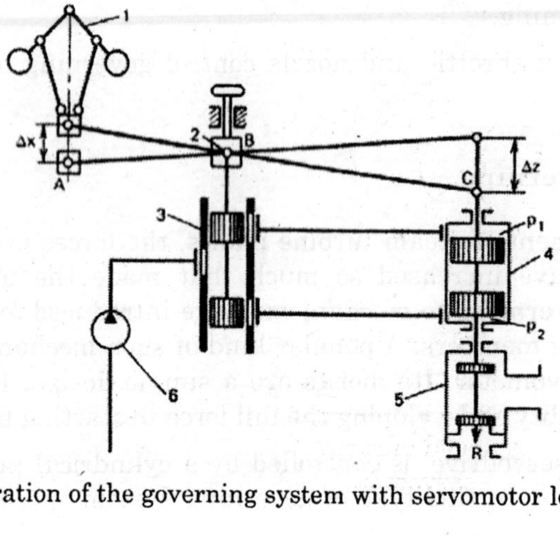

Fig. 4.2. Operation of the governing system with servomotor load shedding.

Beginning from the instant t_1 at which the moment of internal resistance becomes greater than the angular velocity of the turbine rotor begins decreasing.

Suppose now that the equally $\Delta y = 0$ is established at a certain instant t_3. Since the steam valve is still closed, angular velocity continues decreasing. On a downward movement of the pilot valve, liquid will flow into the lower space of the servomotor, the piston starts opening the steam valve, and steam admission to the turbine is renewed. At the instant t_4, the equality is established. At the same time, since the angular velocity is still less than ω_0, the piston will go on moving upwards and angular velocity will start increasing.

An oscillatory transient process would be avoided if the pilot valve come to the mid position at the instant t_1 (see dash curve $\Delta y = f(t)$. This would establish the equality and the servomotor piston would be locked. Since, however, the angular velocity of the rotor is greater than ω_0 at that time, the servomotor sleeve can by no means return into the initial position. This suggests the conclusion that the pilot valve can return into the mid position only under the provision that its movement is determined not only but the movement of the speed governor sleeve, but also by the movement of the servomotor piston. The diagram of such velocity governing is shown in Fig. 4.3. In this case, lever ABC serves as feedback for the servomotor piston.

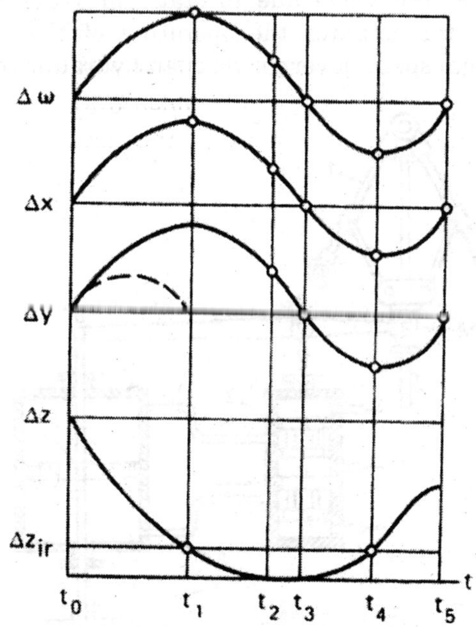

Fig. 4.3. Governing system with feedback and turbine control mechanism.

Comparing the schemes in Fig. 4.1 and 4.3, it may be noticed that in the latter case when the servomotor piston moves in the direction opposite to the movement of the governor sleeve, it decreases the travel of the pilot valve, which reduces the flow of liquid into the servomotor space and, as a consequence, decreases the speed of the piston, which has an adverse effect on the speed of response of the system. An essential property of the feedback is that it retains the servomotor piston in locked position. Suppose that, with a steady state position of the system, the servomotor piston starts moving downwards occasionally (say, due to leakage of the liquid through the gap between the piston and cylinder). The feedback will then cause pilot valve (Fig. 4.3) to move downwards, the liquid will be supplied into the lower space of the servomotor, and the piston will return to the initial position. Thus, the provision of feedback makes it possible to move the servomotor piston rather accurately on variations of static conditions by an impulse of the speed governor.

4.2.2 Throttle governing

The object of throttle governing is to throttle the steam to a suitable pressure and to reduce the steam flow i.e. to allow required quantity of steam to flow through the turbine blading whenever there is a reduction of load compared to economic or design load for maintaining the speed of the turbine.

Fig. 4.4 shows the schematic diagram of throttle governing employing mechanical speed governor (centrifugal governor) and servomotor. Turbine steam stop valve is used to start and stop the turbine by opening and closing it by means of a hand wheel or automatic control. The throttle valve throttles the steam. Various form of throttle valve are used but in this case a double beat valve having seats of nearly equal area and is so shaped that the forces on the valve due to static pressure and dynamic action are balanced. An oil servomotor actuates the operating of throttle valve which in turn controlled by the centrifugal speed governor through hydraulic relay system.

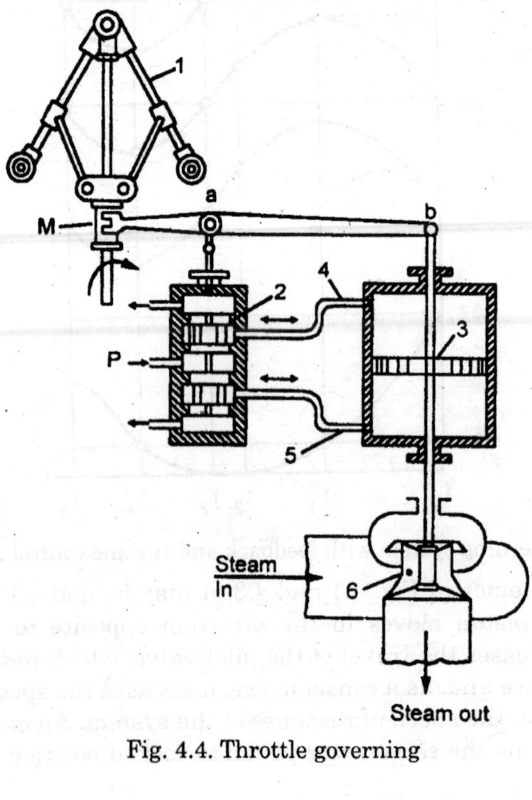

Fig. 4.4. Throttle governing

The change in available enthalpy for throttle governing and Willan's line as is shown in Fig. 4.5.

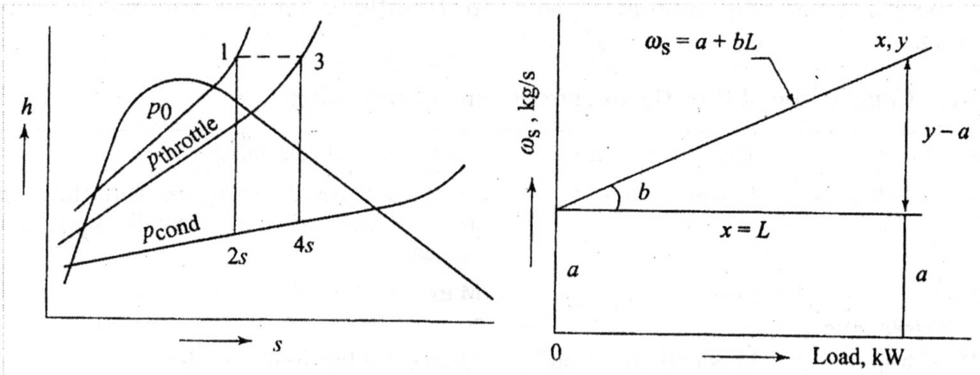

Fig. 4.5. h-s diagram and Willan's line.

4.1.3 Nozzle control governing

In the nozzle control governing, the nozzles of the turbines are grouped in two, three or more groups, and each group of nozzles is fed with steam controlled valves. Various arrangements of valves and group of nozzles are employed. Whatever arrangement is employed, the nozzle control is necessarily restricted to the first stage of the turbine, the nozzle areas in the other stages remaining constant. If the condition of the steam at inlet to the second stage is not materially affected by the changed condition in the first stage, the absolute pressure of steam in front of the second stage nozzles will directly proportional to the rate of steam flow through the turbine.

Fig. 4.6 shows an arrangement often employed in large steam turbines using high pressure steam.

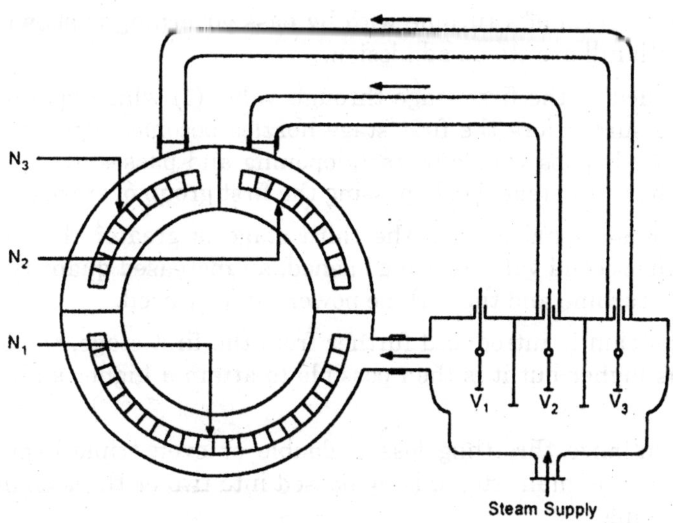

Fig. 4.6. Nozzle control governing

The nozzles are divided into groups N_1, N_2 and N_3 under the control valves V_1, V_2 and V_3 respectively. The number of nozzle groups may vary from three to five or more.

4.1.4 Comparison of throttle governing and nozzle control governing

Table 4.1 gives the differences between throttle governing and nozzle control governing.

Table 4.1. Comparison of throttle and nozzle control governing.

SN	Points	Throttle governing	Nozzle control governing
1	Throttle Losses	Losses are more	Very small throttling losses. Throttling losses are restricted to only partially open nozzle valves.
2	Partial admission loss	Less	More
3	Heat drop available for work	Less enthalpy drop available	More enthalpy drop available.
4	Efficiency	Poor efficiency suitable for small turbines	Suitable for medium and large turbines if initial stages are impulse.
5	Uses	Employed in both impulse and reaction turbines	Employed only in impulse turbines.

4.1.5 By-pass governing

The principle of by-pass governing is to by pass some extra quantity of steam to the far down steam stages when the load is more than economical load.

By pass governing is most often employed, together with throttle governing. In some case, however, it is combined with nozzle governing, in particular, when it is needed to ensure turbine overloading above the economic power.

The principal diagram of a turbine with by pass governing is shown in Fig. 4.7. All stages are made with full are steam admission.

Steam is supplied to the first stage through valve (1) which operates as a throttle valve until the pressure before the first stage nozzles becomes equal to the pressure of the main steam. The bypass valve (2) starts opening and passes over part of the steam flow directly into the third stage, i.e. bypassing the first group of stages.

The clear cross sectional area of the third stage is greater than that of the first stage, so that, as the second valve is being opened, an increased quantity of steam can be passed through the turbine and the turbine power can be raised.

If the bypass steam is introduced further from the first stage, the loss of efficiency due to throttling is higher but it is then possible to attain a higher additional (overload) power.

In order to minimize throttling losses, double or even triple bypass governing is sometimes employed, i.e. main steam is by passed into two or three chambers along the steam path of a turbine.

By pass governing of the type illustrated in Fig. 4.7 is not available for turbines designed for a high initial temperature of steam, because of the following considerations.

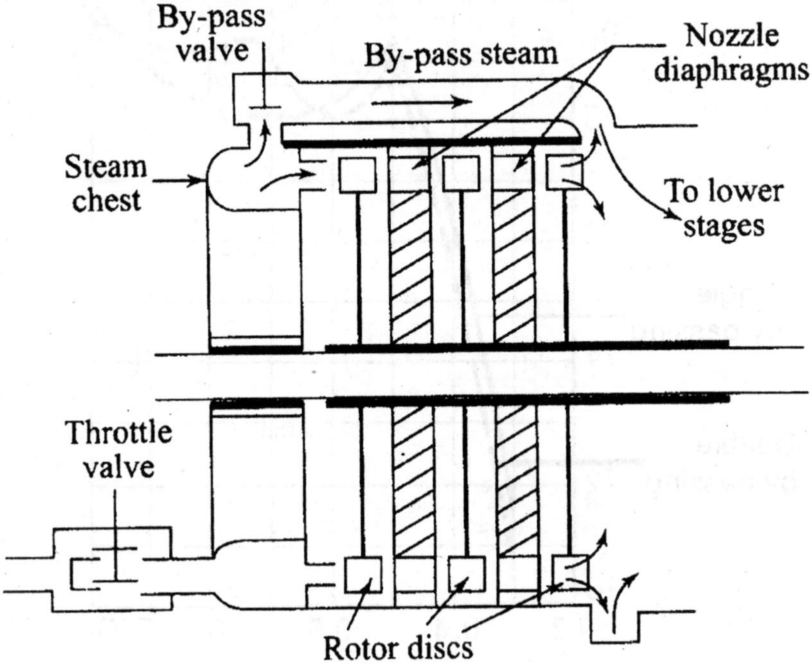

Fig. 4.7. By pass governing.

(1) Firstly, with main steam being supplied into an intermediate stage of a turbine, the group of initial stages and the turbine housing will be subject to the high temperature of main steam, which necessitates the application of high alloyed steels and increases sharply the cost of turbine.

(2) Secondly, since the governing stage works out in the economic regime a relatively low heat drop (most turbines have a single row governing stage) and the pressure in the governing stage chamber is only slightly lower than the initial pressure, it is impossible to attain an appreciable overload rate by opening the governing valve of the additionl group of stages. If the flow rate of steam is increased, it causes a further rise of pressure in the governing stage chamber and reduction of steam discharge through the nozzle segments which have been opened earlier.

In addition, pressure rise in the governing stage is associated with worsening of the stage efficiency and growth of steam temperature in the governing stage chamber above the safe limit. For that reason, turbines designed for a high initial temperature of steam usually have internal (rather than external) steam by passing from the governing stage chamber into an intermediate stage Fig. 4.8.

Variations of the internal relative efficiency depending on steam flow rate for turbines with single and double bypassing are shown in Fig. 4.9. With single bypassing, the turbine overloading begins from m/mo = 0.5 and immediately causes an intensive drop of efficiency due to steam throttling in the bypass valve, which at m/mo = 0.7 may be as high as $\Delta\, \eta_{ri} / \eta_{ri} = 1.85\%$.

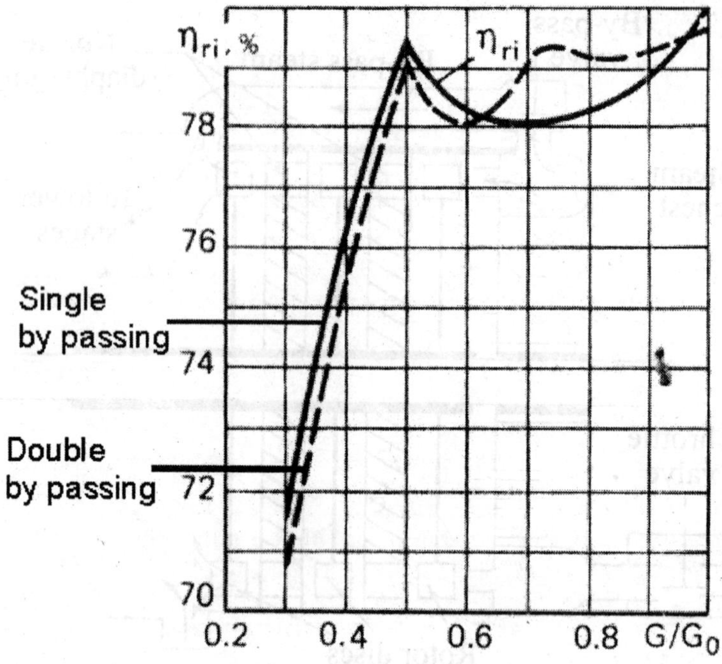

Fig. 4.8. Scheme of steam distribution with internal by passing

With double bypassing after the second bypass valve has been opened fully and does not throttle the steam, the third bypass valve is opened at m/mo = 0.7. Due to this arrangement, the efficiency at m/mo = 0.7 attains the same level as it had before opening the second valve.

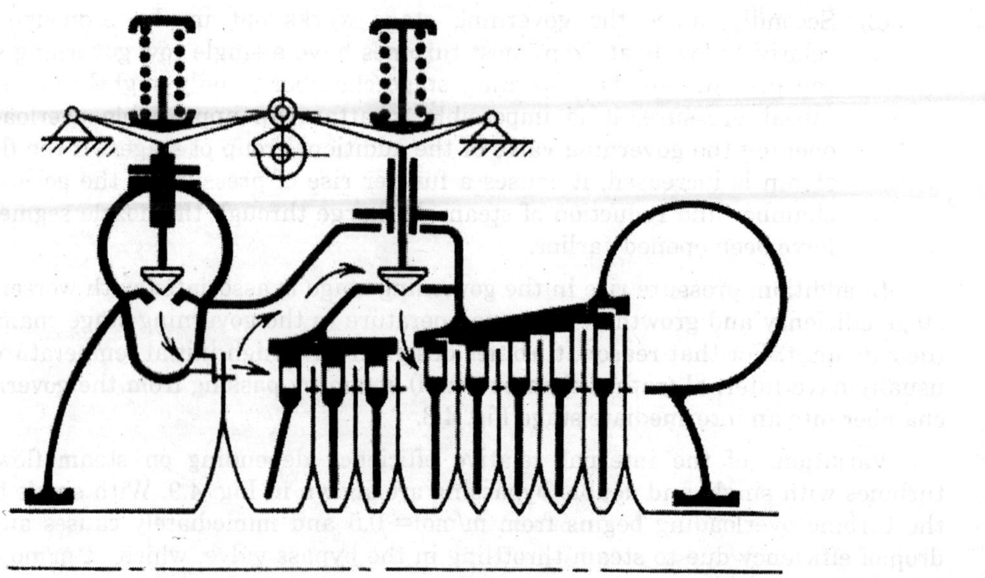

Fig. 4.9. Relative internal efficiency of a steam turbine with bypass governing, depending on steam flow rate.

4.1.6 Selection of steam governing system

The selection of a particular steam governing system is done according to the application of the turbine. If the turbine is to be used for carrying base loads of an electric grid, it should be designed for as high an efficiency as feasible. Such a turbine should be run at the rated power, where possible and can be made with a small number of governing valves and nozzle governing or with purely throttle governing.

The load of modern power grid is, however, subject to strong variations. At many power stations it may drop down more than by 50% at holidays and night dips. During these periods of load dips, a certain relatively small number of turbines is shut down and the remaining continue to operate at a load ranging from 50% to full and at peak hours, at the highest power. For these reasons, most steam turbine plants, including blocs of a capacity of 500 MW and even 800 MW should be designed for operation at substantially reduced (up to 50%) loads, as wel as at full load. Under such conditions, nozzle governing turns out to be the most efficient system, since it worsens the turbine efficiency at partial loads much than does the throttling system.

A circumstance of no less importance is that a change of load of a turbine with nozzle control governing results in intermediate stages, and therefore, of the temperature of turbine casing than with throttle governing. This circumstance is especially important for turbines which have to operate with strong daily variations of the loads associated with frequent shut-down and start-ups of the turbine large and sharp changes of temperature of the turbine casing under such operating conditions can produce temperature gradient and temperature stresses in the turbine casing, which can reduce the manoeuvrability by nd reliability of the turbine. In association with the factors mentioned, steam turbines of very high capacities, especially saturated steam turbines of nuclear power stations, are preferably made with throttle governing.

If the nozzle control governing is employed, it is essential to select properly the optimal available heat drop of the governing state under the rated operating conditions and the optimal number of nozzle chests.

Advantages of nozzle control governing under transient operating conditions are sensible only at a high design heat drop of the governing stage. On the other hand, an increase of enthalpy results in poorer efficiency of the turbine at full load.

Hence, it follows that a lower heat drop of the governing stage should preferably be chosen for turbines operating a long time at full load and a higher heat drop for those whose load is subject to sharp changes.

The problem of selecting the number of nozzle segments of the nozzle control governing system and of the effect of the number of governing valves on the efficiency of turbines at reduced loads is usually solved on the basis of the load curves of turbines. Fig. 4.10 shows the curves of turbine power variation depending on the flow rate of steam through a turbine with purely throttle governing (curve bc) and for an ideal nozzle control governing with an infinite of nozzle group (curve ba). At full load (m/mo=1), the power of both turbines is conditionally taken to be the same (p/po=1), through with throttle governing it can be higher at a higher design steam discharge. With reduction of the steam flow rate, the power of the turbine with throttle governing decreases more than that of the turbine with ideal nozzle control governing. The difference in their powers can be attributed to the losses caused by steam throttling in the admitting valve of the turbine with nozzle control governing and is determined as the difference between the ordinates of the two curves.

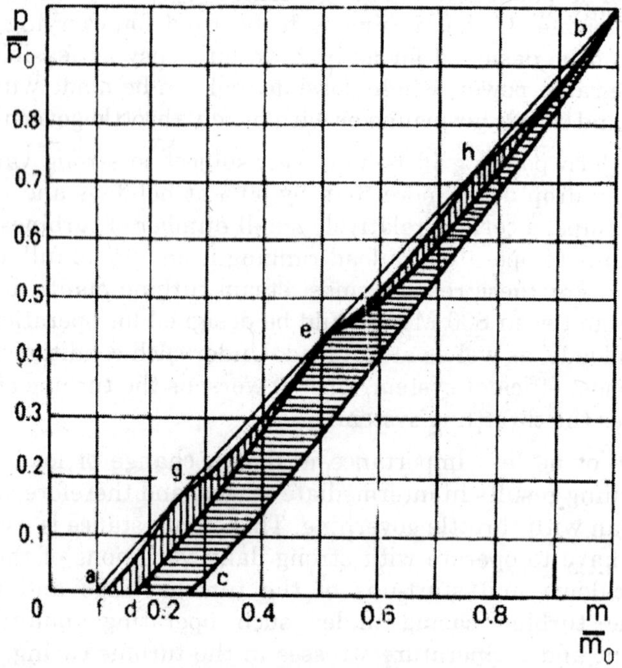

Fig. 4.10. Variations of turbine power depending on changes of steam low rate with various steam governing systems.

In order to estimate the effect of the number of nozze segments on the turbine efficiency at a variable steam discharge, let us divide the nozzle cascade of the first stage into two groups with the same number of nozzle passages and provide each group with an individua governing valve. Then, with one of the valves fully opened, the curve of turbine efficiency will coincide at point *e* with efficiency curve of the turbine with ideal nozzle control governing and the variation of power will be described by line *bed*. The gain in power in the plant with two nozzle groups will be represented by the horizontally hatched area. Now, each of the two groups is against divided into two groups. With four groups of nozzles, the change of power will be described by line *bhegf* and the gain in power compared with the system with two groups of nozzles will be substantially smaller through sensible, and will be represented by the vertically hatched area.

A change from four to a large number of nozzle groups offer quite a small gain in power, but makes the turbine design more intricate. For that reason, nozzle control governing usually have four nozzle groups and very rarely six or eight groups.

The method of connection of nozzle passages into governing groups is selected depending on the expected range of loads of the turbine being designed and on the preferable power at which the turbine will be run the longest time. For example, it is known that the turbine will operate the main time at loads ranging from 0.4 to 0.8 of the maximum power, with short changes to loads less than 0.4 and more than 0.8 of the maximum. Under such conditions, a reasonable solution is that the turbine should develop the power up to 0.4 of the maximum with one governing valve being opened and further increase of the power up to 0.8 of the maximum should be attained by successive opening of the second, third, and fourth valve. Since the turbine will operate at the maximum power only for a short time, this can be effected by arranging a bypass overload valve.

When choosing the sequence of opening of governing valves, it is essential to consider additionally the conditions of heating of the turbine casing. On the basis of these considerations, for instance, two governing valves are often opened simultaneously. One of them supplies steam into the lower half of the cylinder and the other into the upper half, so that both halves are heated up uniformly.

4.2 EMERGENCE TRIPPING OF TURBINE

Modern turbines are accelerated to the rated velocity in 0.3 to 0.35s. With load shedding the rotor speed can change during that time by W_{max}. It should be borne in mind that the spaces between the governing valves and the first row of nozzles contain an appreciable quantity of steam whose flow, even on instantaneous cut off of steam admission to the turbine, can increase additionally the angular velocity. High specific capacities of modern steam turbines preclude the possibility of making sufficiently strong turbine stages, especially the last ones, to withstand a substantially higher rotational velocity which might appear as a result of load shedding. It is therefore essential that the governing system could respond so quickly as to present an increase of rotational frequency by 10 to 11% above the rated valve ω_0.

Probable destructions of a steam turbine after run-away caused by load shedding might be so severe (on a failure of the safety trip) that turbine manufacturers attach the prime importance to their prevention. Steam turbines are provided with independent protection (safety) lines. The first line is the governing system proper and the second is a special safety system. The latter always includes a speed governor which is usually called *overspeed governor* (or safety trip) and an independent steam stop valve which cuts off steam admission to the turbine by the command of the overspeed governor.

4.2.1 Annular overspeed governor

Fig. 4.11 shows the most popular annular overspeed governor.

A relatively massive ring is fitted onto shaft. The ring has a guide pin entering a boring in the shaft. The ring is retained in place by spring which is pressed against the shaft with a force controlled by nut. The ring has two recesses inside, which are filled with oil during turbine inspections. The size of recesses is calculated so that the centrifugal force of oil counter balances the force of the spring and the device is tripped at the rated angular velocity. This allows the safety trip to be tested on a running turbine, for which purpose it is disconnected from the cut off valve actuated by it. Recesses in the ring are filled with oil by means of remote controlled nozzles upon testing of the safety trip and closing the oil supply line, oil is poured from the recesses through small holes. The safety trip is tested after every 4000 hours of turbine operation. The centre of gravity of the ring is shifted by distance r from the axis of rotation.

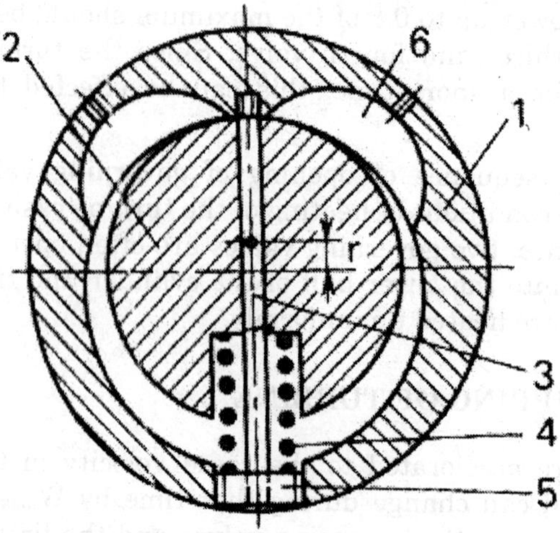

Fig. 4.11. Annular overspeed governor

4.2.2 Annular overspeed governor with side suspension

Fig. 4.12 shows the peculiar design of annular overspeed governor with side suspension. Ring is mounted.

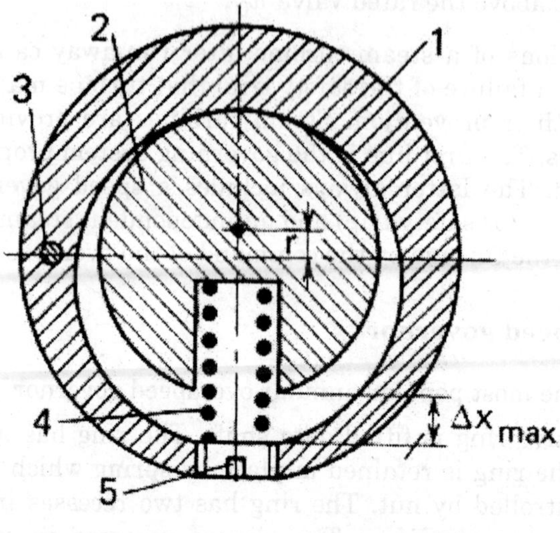

Fig. 4.12. Annular overspeed governor with side suspension

On aile and can tilt on the latter under the action of centrifugal force. As in the previous design, the centre of gravity of the ring is shifted relative to the axis of rotation of the turbine shaft by a distance r. The compression of spring is adjusted by nut. The arm of centrifugal force relative to the tilting axle of the ring exceeds substantially the arm of the action of friction force on the surface of axle, which ensures a high sensitivity of the safety trip. A certain drawback of the design is that the ring plane must be arranged with a high accuracy prependicular to the axis of rotation of the shaft. A skew arrangement can produce a restoring moment which increases frinction on axle.

4.2.3 Cylindrical overspeed governor

Fig. 4.13 shows the cylindrical overspeed governor. It is fastened at the end of turbine shaft and consists essentially of two governors. The centre of gravity of the device is shifted relative to the turbine axis so that the centrifugal force of a striker tends to push out the device againxst the force of a spring. Spring passes the striker to an adjusting nut. The setting of the overspeed governor is controlled by changing the tension of spring. The axes of the two strikers are located so that the centrifugal forces developed on their actuation are direction oppositely and counterbalanced.

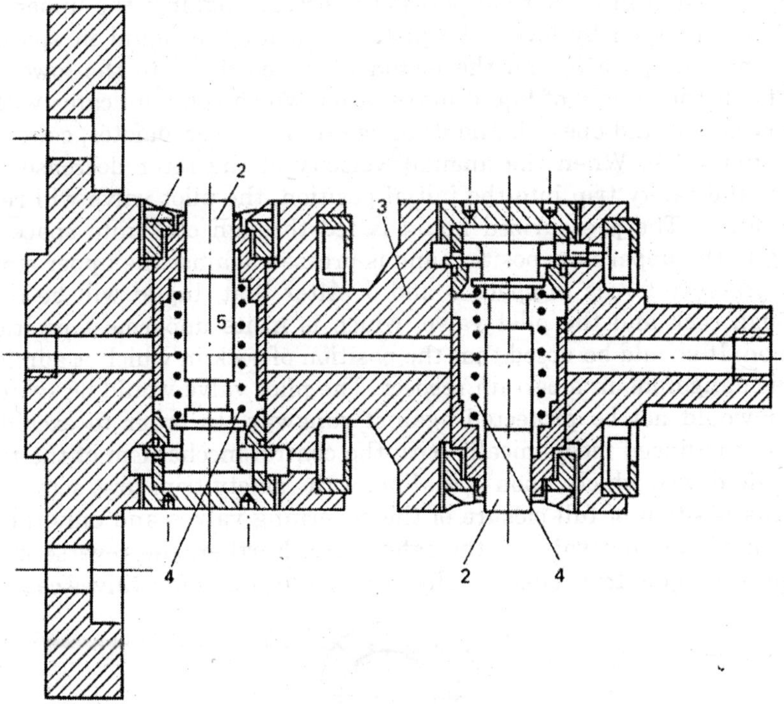

Fig. 4.13. Cylindrical overspeed governor.

The head portion of a striker is essentially a piston supplied with oil from trough shaped grooves. Oil is fed through external pipes and borings under the piston so that combined action of oil pressure and centrifugal force pushes out the piston against the force of the spring. This design prevents seizure of the pistons owing to misalignments.

The springs are compressed by nuts so that the strikers operate on an increase of the angular velocity by 10 to 12% above the rated value.

The design of striker type overspeed governor employ the principle of statically unstable regulator. After operation of the overspeed governor, the striker moves from one extreme position to the other. To return the striker into the initial position, the angular velocity of the rotor must be decreased somewhat below the rated value.

Reliable operation of a safety trip depends to an appreciable extent on the reliability of the mechanism that links the trip with the steam stop valve. A cylindrical pilot valve is widely used as the transmission mechanism. A kind of interlocking should be provided in the latter, particularly, the steam stop valve must be precluded from repeated opening as the angular velocity of turbine rotor increases upon actuation of the safety trip.

4.2.4 Cylindrical pilot valve

Fig. 4.14 shows the cylindrical pilot valve. As the ultimate angular velocity of the turbine rotor is attained, safety trip is deiengaged and strikes level which tilts on pin and depresses the piston of pilot valve. The pilot valve is fed with pressurized liquid which enters the spaces *a*, *b* and *c*. The surface area of piston is space *a* is smaller than that in space *b*, and the piston in its normal position is pressed against the upper stop. A pilot valve has been depressed by lever, the piston area for the liquid in space *a* becomes greater than that in space *b*, and the piston is moved down to the lower stop. In its motion, it cuts off the supply of liquid into space *c* which communicates with the steam stop valve servomotor and opens liquid drainage from the servomotor, causing the latter to close the stop valve. When the angular velocity of the rotor decreases so that the spring returns the safety trip into the initial position, the pilot valve will remain in the lowermost position. The pressurized liquid is fed through cock into space *d* and pilot valve moves into the uppermost position and is pressed against the upper stop due to the difference of pressure forces in space *b* and *a*. After that, liquid will be released from space *d* through cock and the pilot valve, which is in its uppermost position, is again ready for action. It should be noted that the position of cock in which it admits the liquid into space *d* is dangerous in a certain respect. The pilot valve would be switched off and the turbine 1 would not be protected against runaway. In order to exclude fully this probability, cock is lineed mechanically with the control mechanism of the turbine, thus forming a single device which, upon actuation of the safety trip, first sets the governing system into the position of full closure of the governing valves and only then opens the flow of liquid into the pilot valve of the safety trip. Further, the reverse motion of this device releases the liquid from space *d*, after which the governing valves can be opened.

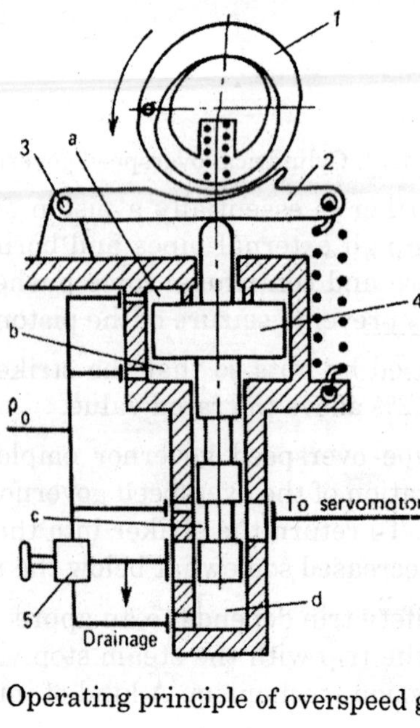

Fig. 4.14. Operating principle of overspeed governor.

To ensure reliable operation of the safety strip, a servomotor with cutoff pilot valve and with the spring acting in the direction of stop valve closing. In this case, the servomotor will close the stop valve as room as the supply of pressurized liquid is disconnected.

To prevent steam shock, a kind of interlocking which prevents the steam stop valve from opening when the governing valves are opened. Such interlocking consists in that the servomotor of the steam stop valve is calculated so that the force acting on its piston is insufficient for opening the steam stop valve under the conditions considered, and therefore, steam shock is avoided.

4.3 COMPARISON BETWEEN IMPULSE AND REACTION TURBINES

The main differences between the impulse and reaction turbines are given in table 4.2.

Table 4.2. Comparison between impulse and reaction turbines.

SN	Points	Impulse turbines	Reaction turbines
1	Blade shape	Profile type and its manufacturing is simple	Aerofoil type and difficult to manufacture
2	Blade passage	Constant cross sectional area since there is no expansion.	Variable cross sectional area (converging type) due to expansion)
3	Number of stages	Because of large pressure drop in nozzles the number of stages are less.	Because of small pressure drop in each stage the number of stages are larger for the same pressure drop. Reaction turbines are multistage turbines only.
4	Pressure drop in nozzles and moving blades	Pressure drop takes place in nozzles and pressure remains constant in moving blades	Pressure drop takes place in both fixed blades (nozzles) and moving blades.
5	Construction	Diaphragm contains the nozzle and the rotor construction is disc or wheel type	Fixed blades attached to casing serve as nozzles and the rotor construction is drum type.
6	Diagram efficiency	Diagram efficiency decreases rapidly with the change in designed blade speed/steam speed ratio.	More working range is available because diagram efficiency v/s blade speed / steam speed into curve is flat for large values.
7	Friction losses v/s Leakage losses	Overall friction losses are more compared to leakage losses	Leakage losses are more compared to friction losses.
8	Space/power	Occupies less space per unit power	Occupies more space per unit power
9	Suitable	For small power	For medium and higher power

4.4 COGENERATION

In a broad sense cogeneration is the production of more than one useful form of energy from the same primary energy source. Cogeneration is an old engineering concept. But a feasible cogeneration system must be technically sound, economically attractive, and environmentally acceptable. The general concept of a cogeneration system is shown in Fig. 4.15.

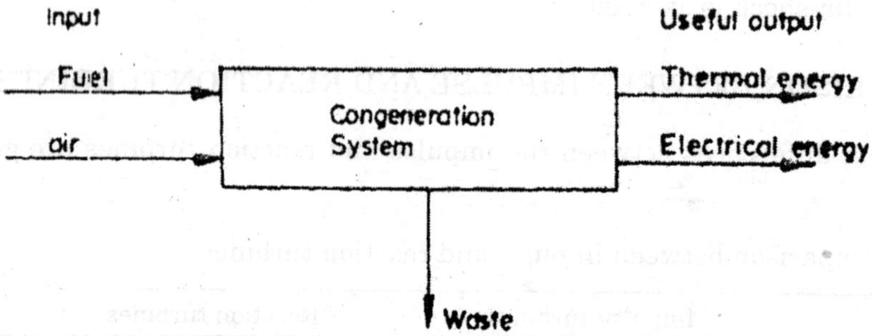

Fig. 4.15 General concept of a cogeneration system

Cogeneration of steam for heating and electric power production is widely used in process industries. Examples include food processing plants, chemical industries, and petroleum production and refining facilities. Industries that use large amounts of process heat also consume a large amount of electric power. Therefore, it makes economical as well as engineering sense to use the already existing work potential to produce power, instead of letting it go to waste.

In reaching the compromising between power and process demands, back pressure and extraction turbines are used.

4.4.1 Back pressure turbine

One simple cogeneration system uses a back pressure turbine, a turbine designed to operate at a high exhaust pressure. In this turbine the pressure drop is through a turbine that delivers power and exhaust at the desired process steam pressure.

A schematic of back pressure turbine system is shown in Fig. 4.16. In this figure no condenser is shown. Thus it is assumed that the process heating load condenses all steam passing through the turbine. The bypass line provides added flexibility for situations where in the process heating load occassionally may exceed the turbine design flow rate. Depending on the desired feed water state leaving the heater, steam may be take from the turbine exhaust to satisfy this requirement.

The primary advantage is the simple system design for cases where little change in power or process steam demand is anticipated. As can be seen, a decrease in power demand will necessitate use of the bypass line to serve a fixed process steam demand. Process steam pressure requirements vary widely, between 0.5 and 40 bar.

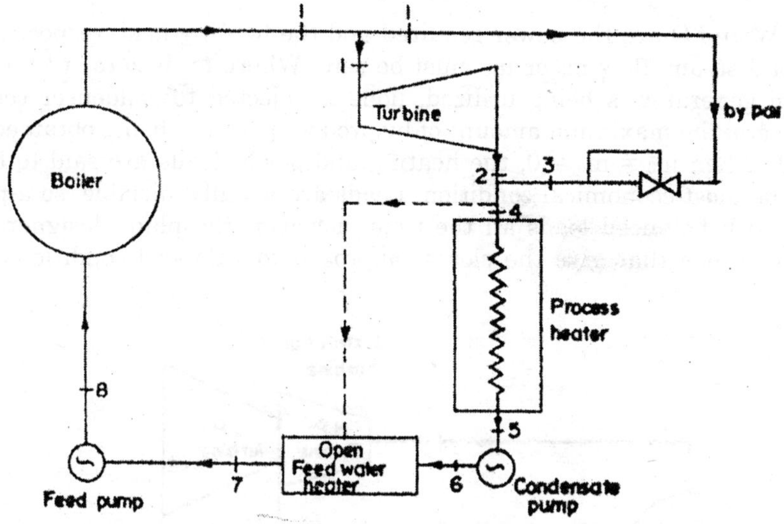

Fig. 4.16. Cogeneration with back pressure turbine

4.4.2 Cogeneration with an extraction turbine (Pass out or bleeder)

A more flexible system for cogeneration results when an extraction turbine is used. The bleed points for feed water heating in power plant turbines are uncontrolled openings in the turbine casing and thus bleed point pressures vary with the power produced. An extraction turbine, in contrast, is one where the steam pressure at the extraction point is controlled even though the turbine load may vary. This is accomplished by feedback control from the extraction point to the turbine inlet valves. This feature results in a higher first cost for the turbine, which is offset by greater flexibility of the system.

A cogeneration system with an extraction turbine is shown in Fig. 4.17. When the power load is zero, all the steam passes through the pressure reducing valve, and non through the turbine. When the heating load is zero, all the steam expands through the turbine and into the condenser. For meeting both heating and power demands, the steam flow distribution is determined as follows: Assuming that all steam to the heating system comes from the turbine extraction point (state 2), the flow rate in the extraction line is determined from an energy balance on the heating system as

$$Q_{heating} = \dot{m}_2 \left(h_2 - h_7\right)$$

Then the power produced by this system flowing through the turbine from throttle to the extraction point is determined from an energy balance on that part of the turbine

$$W = \dot{m}_2 \left(h_1 - h_2\right)$$

The steam distribution problem is to determine the values of m_3, m_3 and m_6 such that their sum is a minimum and they satisfy the first law equations.

$$W = m_2 \left(h_1 - h_2\right) + m_3 \left(h_1 - h_3\right)$$

$$Q = m_2 \left(h_1 - h_2\right) + m_6 \left(h_6 - h_7\right)$$

Where, W and Q are the power required and the heating load, respectively. For the minimum total steam, flow m_3 or m_6 must be zero, Where m_3 is zero, all the heat added in the steam generator is being utilized, none is rejected to condenser cooling water. When m_6 is zero, the maximum amount of by product power is being obtained for a given heating load. When $m_3 = m_6 = 0$, the heating and power loads are said to be balanced, and this is the most economical condition. Loads are usually variable, so a plant cannot be operated with balanced loads all the time, however, the plant designer must select operating conditions that give the closest approach to balanced conditions in the long run.

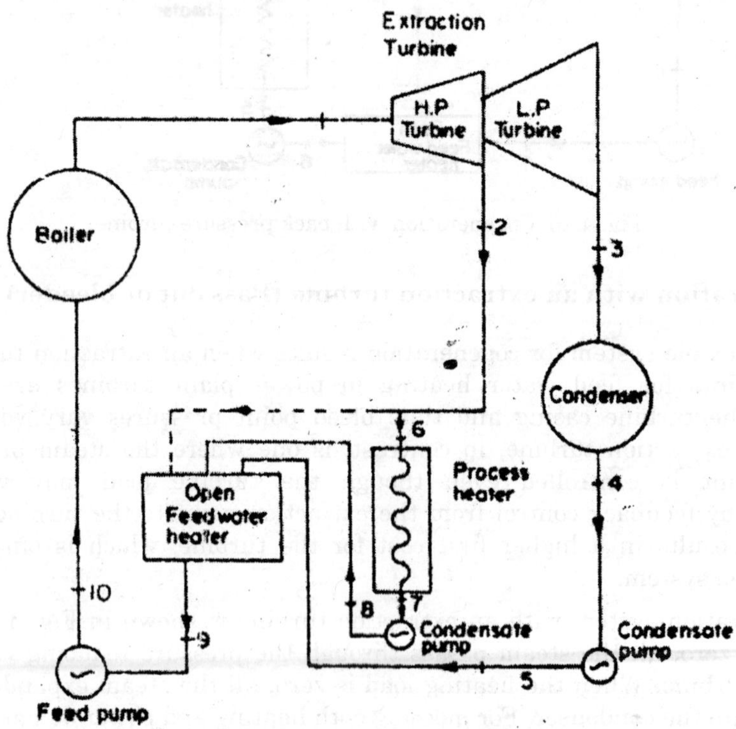

Fig. 4.17. Cogeneration with an extraction turbine

4.5 MIXED PRESSURE TURBINES

This is a turbine which receives steam at more than one pressure, the supplies being fed into the turbine at the appropriate points (Fig. 4.18).

Mixed pressure turbine utilizes high pressure steam from boiler and also admit low pressure steam from the exhaust of a non condensing engine or auxiliaries of the plant. In these turbines governor control is provided to utilize the whole of the exhaust steam available, without allowing exhaust steam pressure to fall and to supply high pressure steam as and when required to keep the speed constant irrespective of load. Depending upon the supply of high pressure and low pressure steam the turbine may work either by high pressure steam alone, by low pressure steam alone or partly by high pressure and partly by low pressure steam.

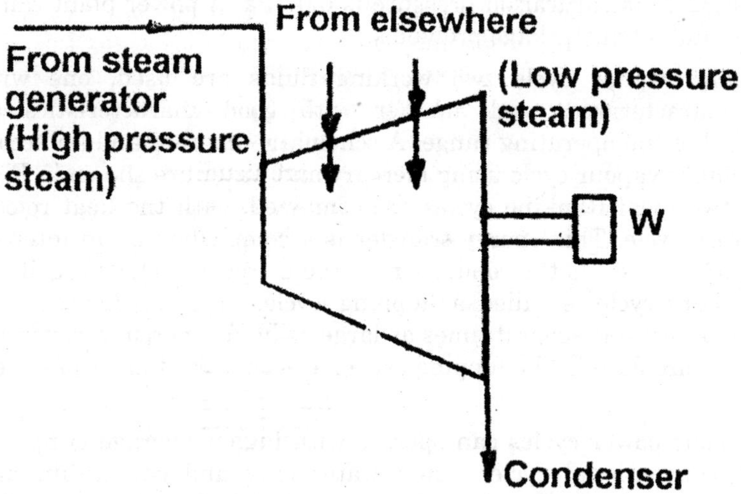

Fig. 4.18. Mixed pressure turbine

4.6 BINARY VAPOUR CYCLE

Water is used as the working fluid in the vast mazority of vapour power systems because it is plentiful and low in cost, nontoxic, chemical stable, and relative non corrosive. In addition, water has a relatively large change in specific enthalpy when it vaporizes at ordinary steam generator pressure, which tends to limit the mss flow rate for a desired power plant output. The properties of liquid water and water vapour are also such that the back work ratios achieved are characteristically quite low, and techniques of reheat and regeneration are effective for increasing power plant efficiencies.

Water has a low critical temperature (374°C, well below the 620°C limit) and very high saturation pressures at high temperatures (165 bar and 350°C). Accordingly, to achieve a high average temperature of heat addition and realize the attentdant higher thermal efficiency, it may be necessary for the steam generator to operate at supercritical pressures. This requires costly piping and heat exchanger tubes capable of withstanding great stresses. Another undesirable characteristic of water is that its saturation pressure at normal condenser temperature is will below atmospheric pressure. As a result, air can leak into the system, necessiating the use of special ejector pumps attached to the condenser or dearating feedwater heater to remove the air.

Cycles that operate at relatively low temperatures may performs best with a refrigerant such as ammonia as the working fluid. Power systems for space flight applications may employ a substance such as mercury as the working fluid because it has desirable performance characteristics at the relatively high temperatures at which these power systems operate.

Working fluids found suitable for the high temperature cycle are mercury, sodium, potassium, and sodium potassium mixtures. The critical temperature of mercury is 898°C (well above the 620° metallurgical limit), and its critical pressure is only about 180 bar. This makes mercury a very suitable working fluid for the topping cycle. Mercury is not suitable as the sole working fluid for the entire cycle, however, since at a condenser temperature of 32°C its saturation pressure is 0.07 Pa. A power plant cannot operate at this vaccum because of air leakage problems.

In a binary vapour cycle two working fluids are used, one with good high temperature characteristics and another with good characteristics at the lower temperature end of the operating range. A schematic diagram and an accompanying T-s diagram of a binary vapour cycle using mercury and water are shown in Fig. 4.19. In this arrangement, two ideal Rankine cycles are combined, with the heat rejection from the high temperature cycle. This energy transfer is accomplished in an interconnected heat exchanger, which serves as the condenser for the mercury cycle and boiler for the water cycle. The mercury cycle is called a `topping' cycle. Since h_{fg} for water at the steam generation temperature is several times as large as h_{fg} for mercury, several units of mass of mercury must circulate in the topping cycle for each unit of mass of water in the other cycle.

Binary vapour power cycles can operate with higher average temperatures of heat addition than conventional cycles using water only and can attain higher thermal efficiencies. However, the savings in fuel costs that might be achieved in this manner are for the present offset by increased costs related to the construction and operation of the more complex cycle arrangement.

Inspection of T-s diagram of Fig. 4.19(b) shows that the binary vapour cycle approaches a Carnot cycle more closely than a steam cycle for the save temperature limits. Consequently higher thermal efficiency can be reached.

The disadvantages of using mercury for binary vapour plants are its toxicity, attacks many metals of common use, does not wet the surface of the metal it contacts and high cost.

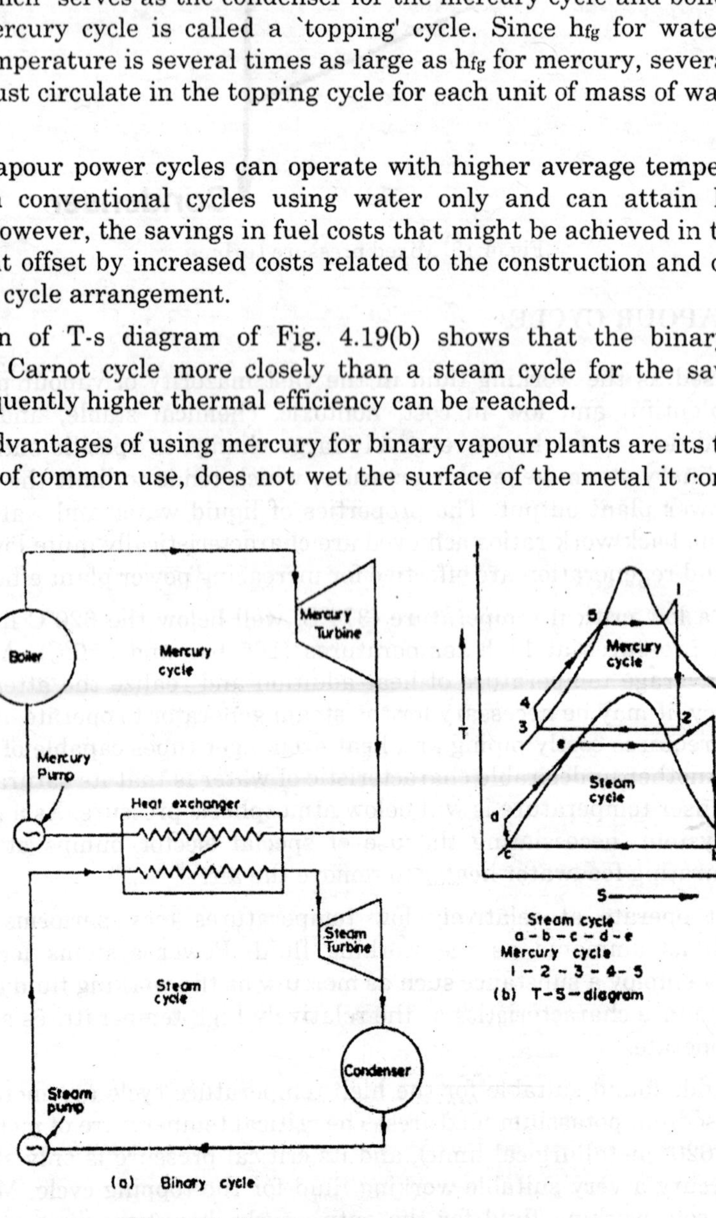

Fig. 4.19. Mercury water binary vapour cycle

ILLUSTRATIVE EXAMPLES

4.1 In a binary vapour power cycle using steam and mercury, saturated mercury vapour enters the mercury turbine at 1.25 *MPa* and exhausts at 14 *kPa*. The condenser steam generator produces a saturated steam at 38 bar. The steam turbine exhausts at 7 *kPa*. Determine (a) the mercury turbine work per kilogram mercury, (b) the steam turbine work per kilogram steam (c) if the steam flow rate is 25 *kg/s*, find the mercury flow rate and (d) the thermal efficiency for the total cycle.

Solution :

p	h_f	h_{fg}	h_g	s_f	s_{fg}	s_g
1.25 Mpa	–	–	361.2	–	–	0.5024
14 kpa	36	294.2	–	0.0923	0.5491	–

For mercury

$h_1 = 361.2\ kJ/kg$ $\quad$ $s_1 = 0.5024\ kJ/kgK$

Using $s_1 = s_2$ $\quad h_2 = 255.7\ kJ/kg$ $\quad$ $h_3 = h_{f3} = 36\ kJ/kg$

For water using the analysis

$h_a = 2802.4\ kJ/kg$ $\quad$ $h_b = 1890.5\ kJ/kg$

$h_c = 163.4\ kJ/kg$ $\quad$ $h_d = 167.2\ kJ/kg$

(a) $W_T = h_1 - h_2$

$= 361.2 - 255.7 = 105.5\ kJ/kg$

(b) $W_T = h_a - h_b$

$= 2802.4 - 1890.5 = 911.9\ kJ/kg$ Ans.

$W_p = 3.8\ kJ/kg$

$W_{net} = W_T - W_P$

$= 911.9 - 3.8 = 908.1\ kJ/kg$ Ans.

(c) $M_{\text{mercury}}(h_2 - h_3) = m_{steam}(h_a - h_b)$

$m_m(255.7 - 36) = 25(2802.4 - 167.2) = m_m = 299.9\ kg/s$ Ans.

(d)
$$\eta_{th} = \frac{m_s(908.1) + m_m(105.5)}{m_m(361.2 - 36)}$$

$$= \frac{25 \times 908.1 - 299.9 \times 105.6}{299.9 \times 325.2} = 0.557 = 55.7\%$$ Ans.

4.2 A binary vapour cycle operates on mercury and steam. Saturated mercury vapour enters the mercury turbine at 450 *kPa* and exhausts at 4 *kPa*. The mercury condenser generates saturated steam at 1500 *kPa* which is expanded in a steam turbine to 4 *kPa*. (a) Find the overall efficiency of the cycle. (b) If 15 *kg/s* of steam flows through the steam turbine, what is the flow through the mercury turbine? (c) Assuming that all processes are reversible what is the useful work done in the binary vapour cycle for the specified steam flow? (d) If the steam leaving the mercury condenser is superheated to a temperature of 300°C in a superheater located in the mercury boiler and if the isentropic efficiencies of the mercury and steam turbines are 85% and 87% respectively, calculate the overall efficiency of the cycle. The properties of saturated mercury are given below.

p kPa	t°C	h_f	h_g	s_f	s_g	v_f	v_g
		kJ/ kg		kJ/kgK		m³/kg	
450	450	62.93	355.98	0.1352	0.5397	79.9×10⁻⁶	0.068
4	216.9	29.98	329.85	0.0808	0.6925	76.5×10⁻⁶	5.178

Solution :

(a) $h_1 = 355.98\ kJ/kg$ Using $s_1 = s_2 x_2 = 0.75$ $h_2 = 254.88\ kJ/kg$

$$W_T = h_1 - h_2 = 355.98 - 254.88 = 101.1\ kJ/kg$$

$$W_p = h_4 - h_3 = v_{g2}(p_1 - p_2)$$

$$= 76.5 \times 10^{-6} \times 446 = 3.41 \times 10^{-2}\ kJ/kg$$

$$W_{net} = W_T - W_P = 101.1 - 341 \times 10^{-2} = 101.1\ kJ/kg$$

$$Q = h_1 - h_4 = 355.98 - 29.98 = 326\ kJ/kg$$

$$\eta_{mercury} = \frac{W_{net}}{Q} = \frac{101.1}{326} = 0.31 = 31\%$$

For the steam cycle

$h_a = 2792.2\ kJ/kg$

$s_a = s_b,$ $x_b = 0.748,$ $h_b = 1941.27\ kJ/kg$

$$W_T = h_a - h_b = 2792.2 - 1941.27 = 850.93\ kJ/kg$$

$$W_p = h_a - h_{fc} = 0.001 \times 14.96 \times 100 = 1.496\ kJ/kg$$

$$h_d = 121.46 + 1.496 = 122.956\ kJ/kg$$

$$Q = h_1 - h_4 = 2792.2 - 122.956 = 2669.24\ kJ/kg$$

$$W_{net} = W_T - W_P = 850.93 - 1.496 = 849.434\ kJ/kg$$

$$\eta_{steam} = \frac{W_{net}}{Q} = \frac{849.434}{2669.24} = 0.318 = 31.8\%$$

$$\eta_{overall} = \eta_{mercury} + \eta_{steam} - \eta_{mercury} - \eta_{steam}$$

$$= 0.31 + 0.318 - 0.31 \times 0.318 = 0.628 - 0.09858$$

$= 0.52942 = 52.942\ \%$ Ans.

(b) $m_m(h_2 - h_3) = m_s(h_a - h_d)$

$$m_m = \frac{m_s(h_a - h_b)}{(h_2 - h_3)} = \frac{15(2669.24)}{254.88 - 29.98}$$

$= 15\times 11.87 = 178.05\ kg/s$ Ans.

(c) $(W_T) = m_m(h_1 - h_2) + m_s(h_a - h_b)$

$= 178.05 \times 101.1 + 15 \times 850.93 = 18000.855 + 12763.95$

$= 30764.805\ kW = 30.76\ MW$ Ans.

(d) $h_1 - h_{2'} = \eta_t(h_1 - h_2) = 0.85\times 101.1$

$= h_{s'} = 270.04\ kJ/kg$

$= m_{m'} = (h_{2'} - h_3) = m_s(h_a - h_d)$

$$= m_{m'} = \frac{15\times 2669.24}{240.06} 166.7858\ kg/s$$

$Q = m_{m'}(h_1 - h_4) + m_s(h_a - h_a)$

$= 166.7858 \times 326 + 15\ (3037.3 - 2792.2)$

$= 54372.171 + 15 \times 281.1 = 58588.671\ kJ/s$

Equating $s_{a'} = s_{b'}$

$x_{b'} = 0.806$

$h_{b'} = 2082.38\ kJ/kg$

$h_a - h_{b'} = \eta(h_a - h_{b'})$

$= 0.87\ (3037.3 - 2082.38) = 0.87 \times 954.92 = 830.78\ kJ/kg$

$W_T = 166.7858\times 85.94 + 15\times 830.78$

$= 14333.572 + 12461.7 = 26795.72$

$$\eta_{overall} = \frac{W_T}{Q} = \frac{26795.272}{58588.671} = 0.4572 = 45.73\%$$ Ans.

EXERCISES

4.1 What do you mean governing of steam turbine.

4.2 Draw a neat sketch of servomotor governing and explain.

4.3 Draw and explain throttle governing.

4.4 Draw and explain nozzle control governing.

4.5 Draw and explain byepass governing.

4.6 How will you select a best governing system on changes of steam flow rate and steam power output and compare them.

4.7 Compare throttle and nozzle control governing.

4.8 Explain emergence tripping of steam turbine.

4.9 Draw a neat sketch of annular overspeed governor and explain.

4.10 Draw a neat sketch of annular overspeed governor with side suspension and explain.

4.11 Draw a neat sketch of cylindrical overspeed governor.

4.12 Explain the operating principle of overspeed governor.

4.13 Explain the purpose of using cylindrical pilot valve.

4.14 What do you meant by cogeneration.

4.15 Draw a neat sketch of back pressure turbine.

4.16 Draw a neat sketch of pass out turbine.

4.17 Draw a neat sketch of mixed pressure turbine.

4.18 Draw a neat sketch of binary vapour cycle.

4.19 Why is mercury a suitable working fluid for the topping portion of a binary vapour cycle but not for the bottoming cycle?

4.20 Discuss the advantages of binary vapour plants over a single vapour plant. What is the present status of binary plants?

4.21 In mercury water binary cycle pressure loss between the mercury boiler and the turbine produces several degree of superheat. Show this on T-s diagram of mercury cycle.

4.22 Sketch a mixed pressure turbine and explain its working.

4.23 How and where back pressure turbine and pass out turbine are used? Give the schematic layout of such plants.

4.24 What is the difference between cogeneration and regeneration?

Chapter - 5

STEAM POWER PLANTS

5.1 INTRODUCTION

A steam power plant continously converts the energy stored in fossil fuels (coal, oil, natural gas) or fissile fuels (uranium, thorium) into shaft work and ultimately into electricity. The working fluid is water which is sometimes in the liquid phase and sometimes in the vapour phase during its cycle of operations.

Steam power plants can be either non condensing or condensing. In non condensing operation as shown in Fig. 5.1, the steam exhausted from the prime movers is discharged at atmospheric pressure or at pressures greater than atmospheric.

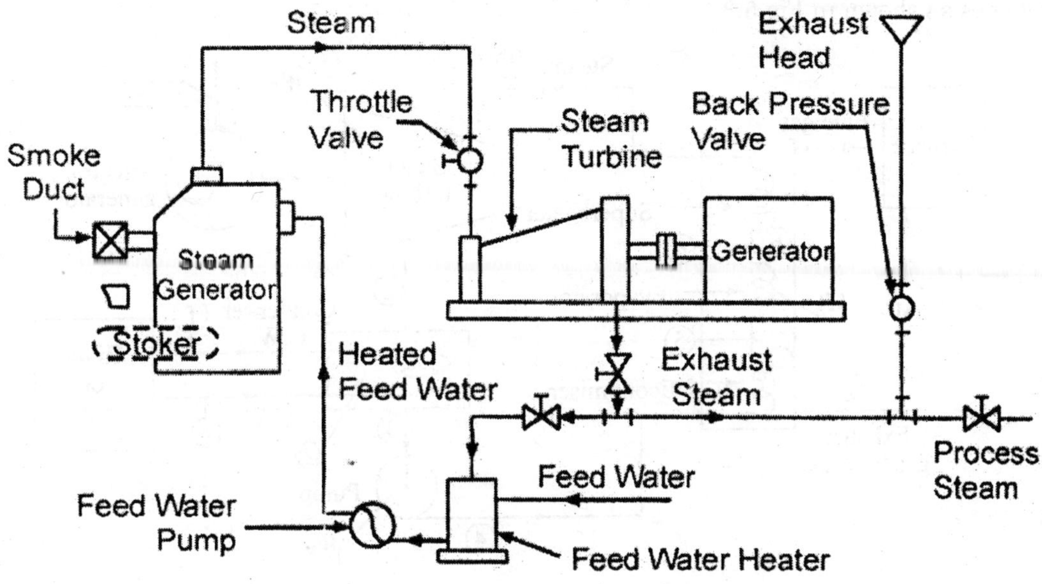

Fig. 5.1 Non condensing steam power plant

In condensing power plants as shown in Fig. 5.2 the exhaust steam is discharged into a condenser instead of into the atmosphere. The pressure in the condenser is less than atmospheric and in it the steam is condensed into water.

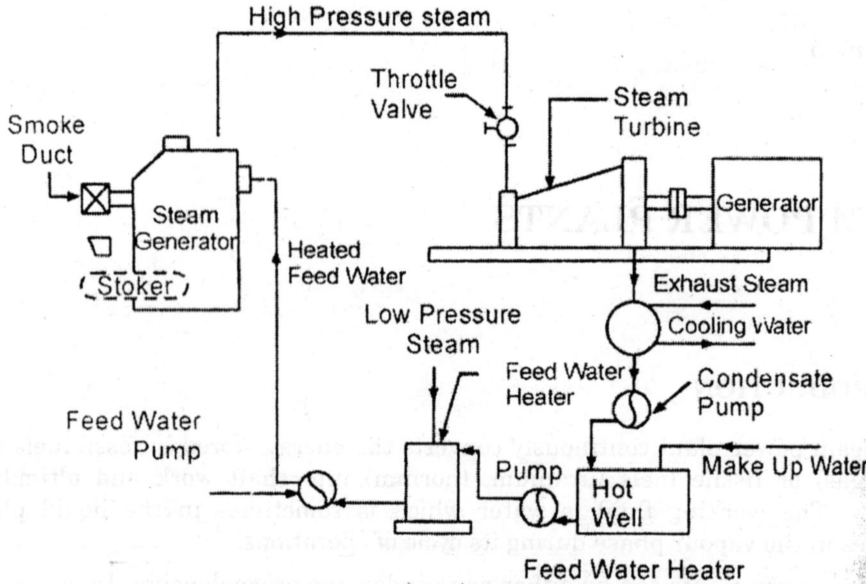

Fig. 5.2 Condensing steam power plants

Heat transfer to water in the steam generator takes place in the three different regimes as shown in Fig 5.3.

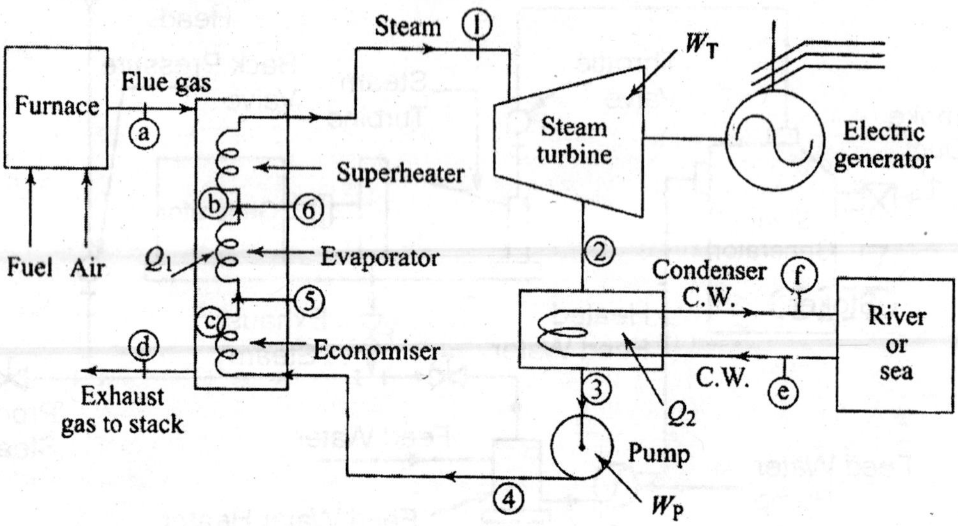

Fig. 5.3 A simple steam plant

5.2 STEAM POWER PLANT LAYOUT

The general layout of the steam power plant consists of mainly four circuits as shown in Fig. 5.4.

1. Coal and ash circuit

2. Air and gas circuit
3. Feed water and steam flow circuit
4. Cooling water circuit.

The brief description of these circuits is given below.

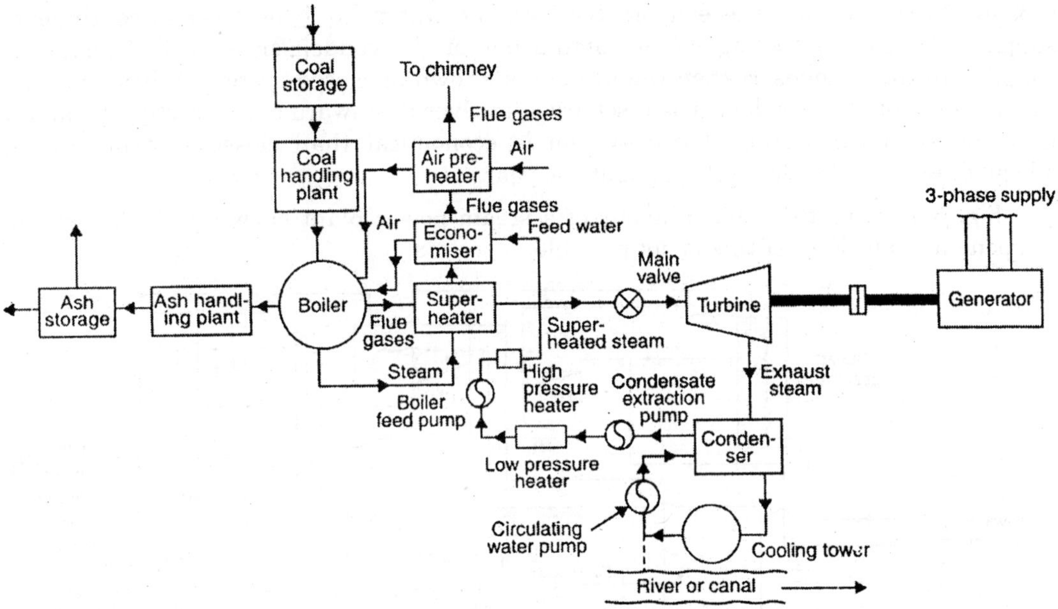

Fig. 5.4 Steam power plant layout

1. Coal and Ash circuit: The coal from the storage is fed to the boiler through coal handling equipment for the generation of steam. Ash produced due to the combustion of coal is removed to ash storage through ash handling system.

2. Air and gas circuit: Air is supplied to the combustion chamber of the boiler through F.D. or I.D. fan or by using both. The dust from the air is removed before supplying to the combustion chamber. The exhaust gases carrying sufficient quantity of heat and ash are passed through the air-heater where the exhaust heat of the gases is given to the air and then it is passed through the dust collectors where most of the dust is removed before exhausting the gases to the atmosphere through chimney.

3. Feed Water and Steam Circuit: The steam generated in the boiler is fed to the steam prime mover to develop the power. The steam coming out of prime mover is condensed in the condenser and then fed. Some of the steam and water is lost passing through different components of the system, therefore, feed water is supplied from external source to compensate this loss. The feed water supplied from external source is passed through the purifying plant to reduce the dissolved salts to an acceptable level. The purification is necessary to avoid the scaling of the boiler tubes.

4. Cooling Water Circuit: The quantity of cooling water required to condense the steam is considerably large and it is taken either from lake, river or sea. The cooling water is taken from the upper side of the river, it is passed through the condenser and heated water is discharged to the lower side of the river. Such system of cooling water supply is possible if adequate cooling water is available throughout the year. This system is known as open system. When the adequate water is not available, then the water coming out from the condenser is cooled either in cooling pond or cooling tower. The cooling is effected by partly evaporating the water. This evaporative loss (this includes evaporation and carryover) is nearly 2 to 5% of the cooling water circulated in the system. To compensate the evaporative loss, the water from the river is continuously supplied. When the cooling water coming out of the condenser is coolied again and supplied to the condenser, then the system is known as closed system. When the water coming out from the condenser is discharged to river downward side directly, the system is known as open system. Open system is economical than closed system provided adequate water is available throughout the year.

Fig. 5.5 shows the typical layour of a steam power plant showing all the relevant components in it. Even if this is not possible.

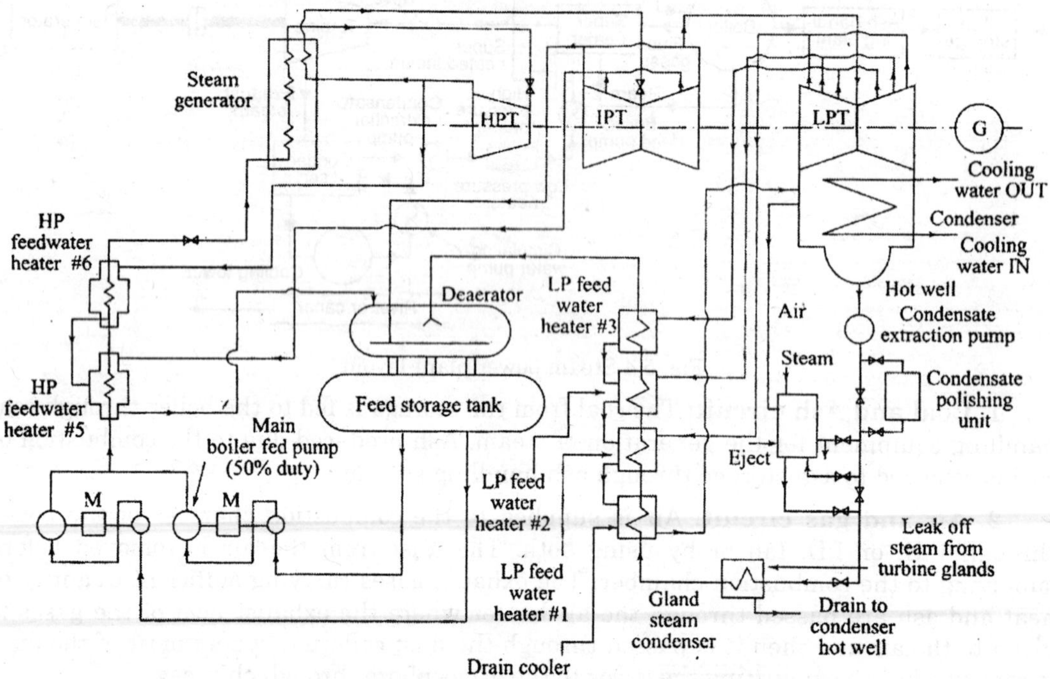

Fig. 5.5 Typical layout of a steam power plant

5.3 SITE SELECTION FOR STEAM POWER PLANTS

The following points should be taken into consideration while selecting the site for a steam power plants.

1. Availability of raw material: A huge quantity of raw material (coal or oil). Therefore, it is necessary to instal the power plant near the coalfields in order to save the transportation cost.

If it is not possible to locate the plant near the coal fields then the plant should be located as near the railway station as possible then at least arrangement should be made for railway siding to the power plant so that the coal wagens can be shunted from the station to the side of power plant. This applies to plants using oil as fuel, as well.

2. Nature of land: The type of the land to the selected should have good bearing capacity as it has to withstand the dead load of the plant and the forces transmitted to the foundation due to the machine operation. The minimum bearing capacity of the land should be 1 MPa.

3. Cost of land: Considerable area is required for the power stations. The cost of the land for that purpose should be reasonable. The large plants in the heat of the big cities and near the load centre are not economical as the cost of land is very high.

4. Availability of Water: Steam power stations use water as the working fluid which is repeatedly evaporated and condensed. Theoretically there should be no loss of water, but infact some make up water is required. Besides this, considerable amount of water is required for condensers. The large quantity of water is also required for disposing the ash if hydraulic system is used. It is, therefore necessary to locate the power plant near the water source which will be able to supply the required quantity of water throughout the year.

5. Transport Facilities: This is another important consideration in locating the thermal power station. It is always necessary to have a railway line available near the power station for bringing in heavy machinery for installation and for bringing the coal.

Another reason in locating the thermal station at pit heads is, there is every possibility of danger arising from dislocation of coal through lock-outs and strikes in the railways. Also in the event of any war emergency, the transport capacity is unavoidably tied up in moving coal for power generation. As against, the modern transmission lines are so constructed that the maintenance required is very less. Therefore, the operating costs of the transmission line can be stated to be practically unaffected by rise in wages and material cost while the same is not true for transfort as evident from the frequent increase in freight rates.

The problem of access during construction is also equally important.

6. Ash disposal facility: The ash handling problems is more serious than coal handling because it comes out in hot condition and it is highly corrosive. Its effect on atmospheric pollution are more serious as the human health is concerned. Therefore, there most be sufficient space to dispose of large quantity of ash.

7. Availability of Labour: Cheap labour should be available at the proposed site as enough labour is required during construction of the plant.

8. Size of the Plant: In small capacity plants, the cost of getting fuel into the plant and ease of water supply are relatively insignificant factors and the problem of plant location reduces almost entirely to an electric transmission problem. Other things being equal, the plant must be located at such a point that the investment in electric cables for transmitting the power together with the annual operating costs should be minimum. If such location takes the plant away from the point of fuel supply or water supply or both, the only saving in the electric transmission may be more than offset by extra costs in the fuel and water supply. Therefore, it is necessary to determine the relative significance of these items before selecting the site for the powerplant. The expenses involved in electric transmission from a small plant are relatively severe owing to the impracticability of using high voltages, so that the electric transmission feature alone becomes dominant in the plant location.

9. Load centre: The power plant should be located, as far as possible near the load centre in order to minimize the cost of transmission lines and also the losses occuring in them.

10. **Public Problems**: The proposed site should be far away from the towns to avoid the nuisance from smoke, fly ash and heat discharged from the power plant.

11. Future extension: The choice of the site should allow for economical extension consistent with estimated growth of load.

5.4 WORKING MEDIUM

The steam power plant needs the following quantities: coal, water, and air.

These are considered the basic needs of the plant. The quantities required are large.

Coal is the oldest fuel and still used on large scale throughout the world for power generation. It is a fuel which brought about an industrial revolution and made a dramatic change in the life style of people. It has sustained most of our combustion activity until oil was struck and even now continues to be a major source of energy.

After the first and second shocks of oil prices, the poer industry in the western countries moved from oil to coal. Mankind has realised that the future lies in establishing an energy economy at the earliest possible time, based on sun, wind, ocean tides and thermal gradients, bioman etc. All effects are goingon throughout the world to develop such non-conventional ever lasting sources of energy. But it will take time, a couple of decades or more to develop these sources and in the present condition of soaring oil prices, the coal is going to play a vital role in the power generation industry.

Keeping this in mind, all oil based power plants are shifting to coal based plants in USA and Europe. In India, coal is the only source of power generation, and it will play very important role in coming years.

In India coal is the primary source of energy and coal India is the controlling body for the coal industry. India has reserves of both oil and coal. While coal reserve in the country are sufficient to last for save hundreds of years. Oil represents only a fraction of total coal reserves.

Indian coals are generally characterised by high ash content and difficult cleaning characteristics. The coals however have a very low sulphur contents. The Indian coal resources constitute only about 0.8% of total coal reserve of the world, India ranking 10th on this basis.

5.5 FUELS AND FUEL HANDLING EQUIPMENTS

Three types of fuels can be burnt in any types of steam generaly plant:

(1) Solid fuel such as coal

(2) Liquid fuel as oil and

(3) Gaseous fuel as gas

Since our country is deficient in petroleum products, the main fuel used in the steam power plants is coal. Coal can be handled manually or mechanically. Mechanical handling is usually adopted due to the following reasons.

(1) No dependence on manual labour.

(2) To minimise handling charges.

(3) Mechanical handling is cheaper in the long run.

(4) Mechanical handling required for unload the coal supply from wagons and barges.

Due to high cost of the coal may be as high as 60 to 70% of the total station operation cost. Of this, nearly half or more is contributed by transportation and handling of coal.

For designing coal handling plant, the following points should be considered.

(1) Reliable.

(2) Simple and sound.

(3) Require minimum operation.

(4) Minimum maintenance.

(5) Avoid double handling.

(6) To avoid abrassiveness and corrosions, working parts should be closed.

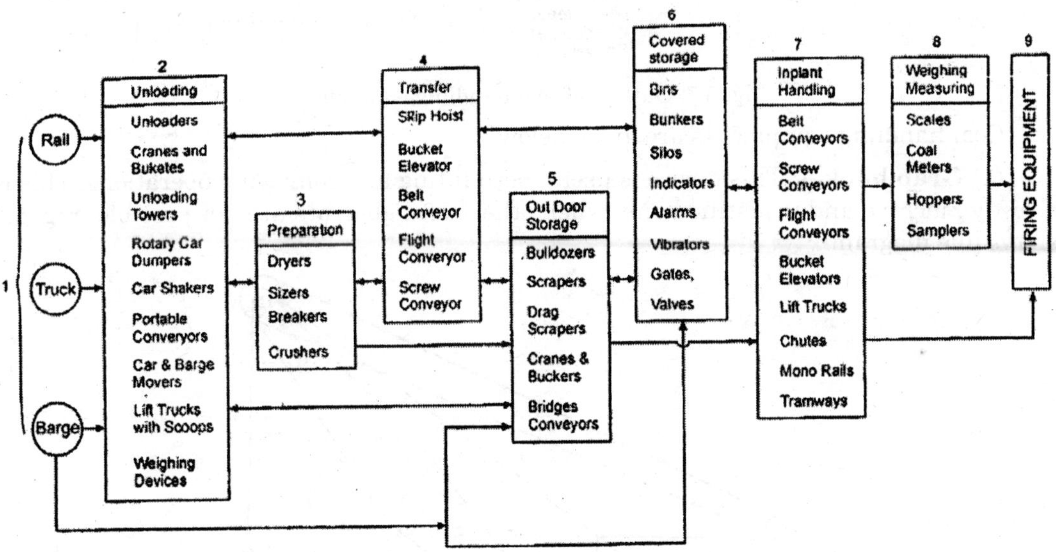

Fig. 5.6 Various stages in coal handling

5.5.1 Methods of coal handling

The following stages/steps are involved in handling of coal.

(1) Coal delivery.

(2) Unloading.

(3) Preparation.

(4) Transfer.

(5) Storage of coal.

(6) In plant handling.

(7) Weighing and measuring.

(8) Furnance firing.

Figs. 5.6 and 5.7 shows various stages and outline of coal handling of equipments.

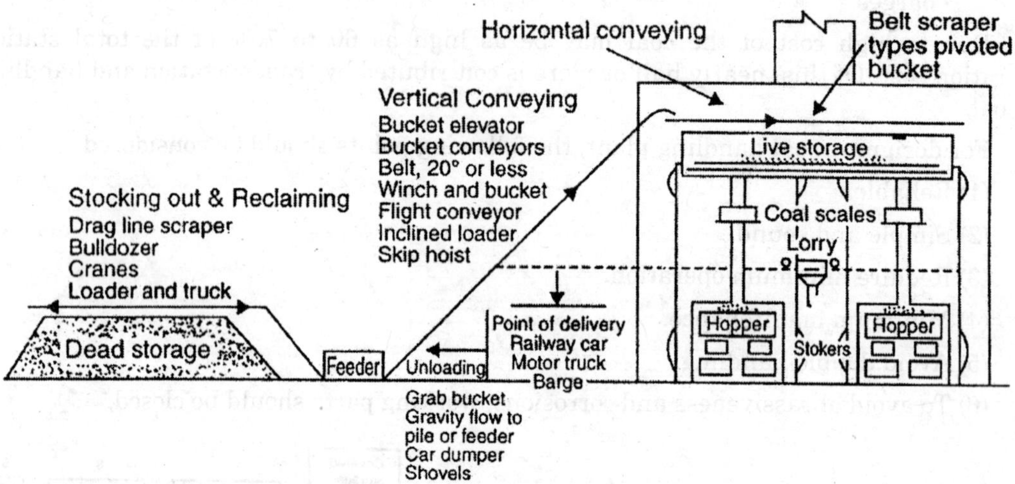

Fig. 5.7. Outline of coal handling equipment.

Coal handling equipments are given below.

(1) Grab bucket: This device is useful both lifting and conveying operations. This is a costly machine and is justified only when other arrangements are not possible. Fig. 5.8 shows line diagram.

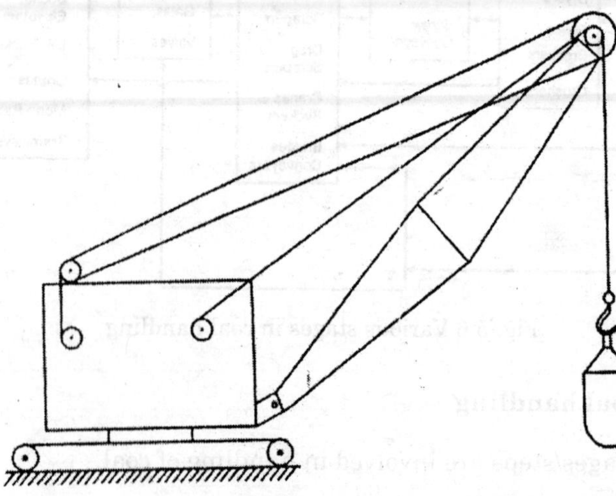

Fig. 5.8 Grab bucket

(2) Bucket elevator: This is used for moderate lifts. The materials is continuously handled and can be both hoisted and conveyed. The speed of the conveyor ranges between 15 m/min to 36 m/min. Fig. 5.9 shows line diagram.

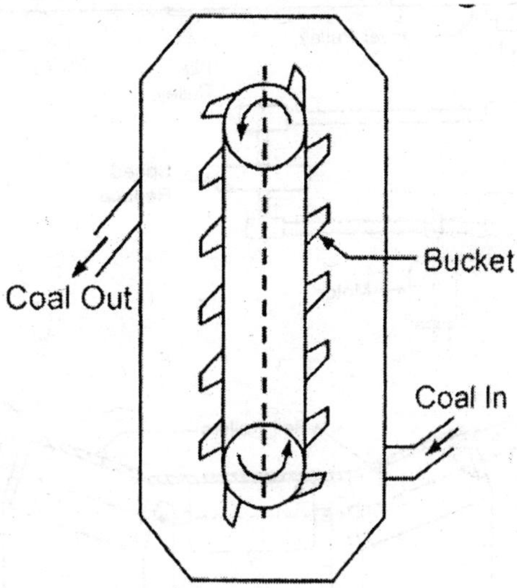

Fig. 5.9 Bucket elevator.

(3) Skip hoist : This is used for high lifts and handling is non continuous. The arrangements is simple and cheap and operation cost including labour and maintenance low. Fig. 5.10 shows line diagram.

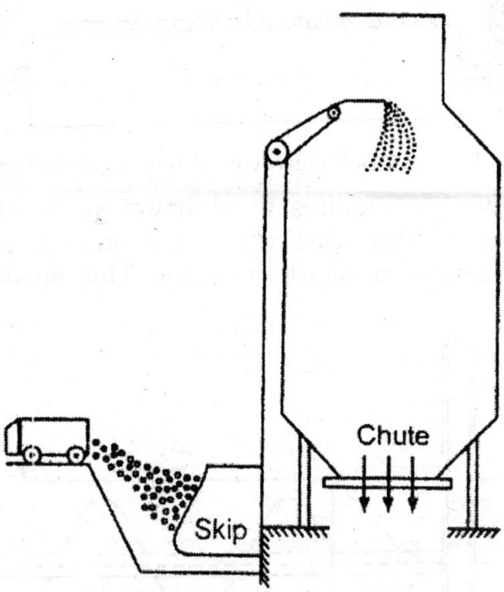

Fig. 5.10 Skip hoist

(4) Belt conveyor : This is a hoisting and conveying device. They are limited to less than 20° to the horizontal, therefore for high lifts the length of the belt is very much increased resulting in high installation cost. Fig. 5.11 shows layout of the belt conveyor and section through the belt conveyor shown.

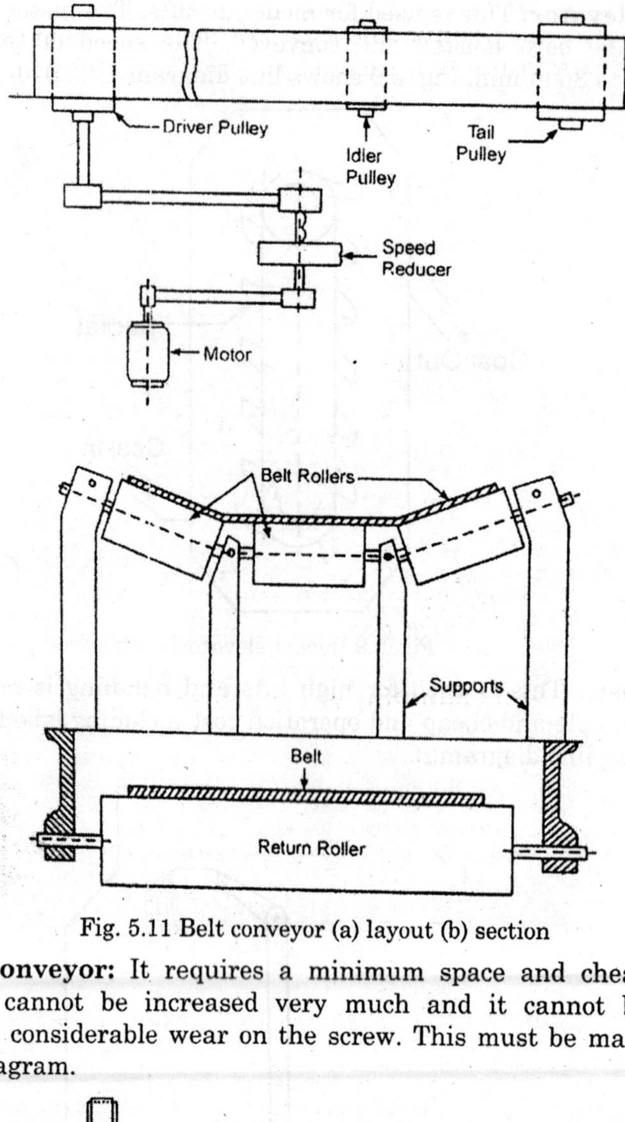

Fig. 5.11 Belt conveyor (a) layout (b) section

(5) Screw conveyor: It requires a minimum space and cheap in installation. However, length cannot be increased very much and it cannot be used for large capacities. In this considerable wear on the screw. This must be made dust tight. Fig. 5.12 shows line diagram.

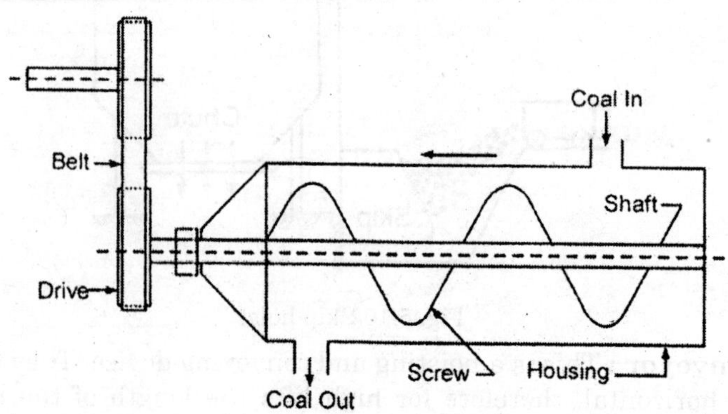

Fig. 5.12. Screw conveyor

(6) Flight or scraper conveyor: It is low in first cost but large energy consumption. There is a considerable wear caused by friction and abrasion. Fig. 5.13 shows line diagram.

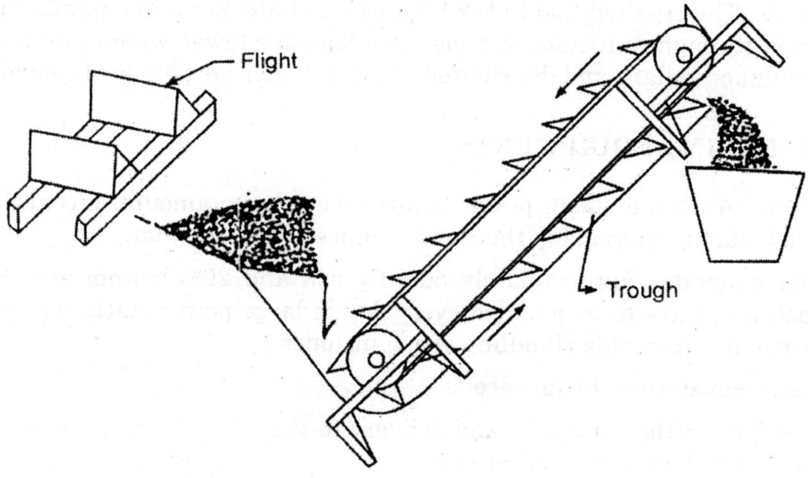

Fig. 5.13 Scraper conveyor

5.5.2 Modern layout of fuel handling equipment

Fig. 5.14 illustrates a typical layout of the modern fuel handling plant. In this plant solid fuel (coal) is used. Coal is delivered to the plant in railway wagons.

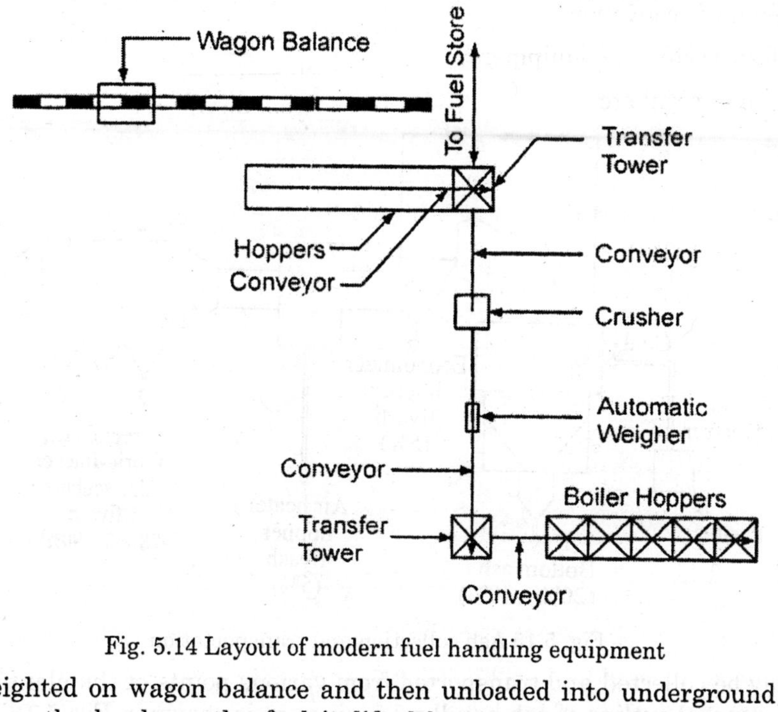

Fig. 5.14 Layout of modern fuel handling equipment

It is weighted on wagon balance and then unloaded into underground hoppers or bunkers. From the bunkers, the fuel is lifted by conveyor to the transfer tower from where it can be delivered either to the fuel store or by a conveyor to crusher. After passing through magnetic separators and screens, fuel is crushed in crusher into pieces 25 to 30 mm in size for stoker firing and 10 to 20 mm when pulverized fuel is fired in boiler furnaces. The crushed coal in the later case is milled to a fine powder and then the fuel is carried through automatic weigher to a transfer tower where fuel is lifted to the level of the bunker gallary and distributed between boiler hoopers by a conveyor.

5.6 ASH HANDLING EQUIPMENTS

In the modern large steam power plants where large amount of coal are used, the amount of ash may go upto many thousands tonnes of ash per year.

A boiler generates approximately 80% fly ash and 20% bottom ash. Hundreds of tonnes of ash may have to be handled every day in large power stations and mechanical devices become indispensible Handling of ash includes

(i) Its removal from the furnace

(ii) Loading on the conveyors and delivery to the fill or dump from where it can be disposed off by sale or otherwise.

The commonly used equipment for ash handling in large and medium size plants may comprise of :

(i) Bucket elevator.

(ii) Bucket conveyor.

(iii) Belt conveyor.

(iv) Pneumatic conveyor.

(v) Hydraulic sluicing equipment.

(vi) Trollies or rail etc.

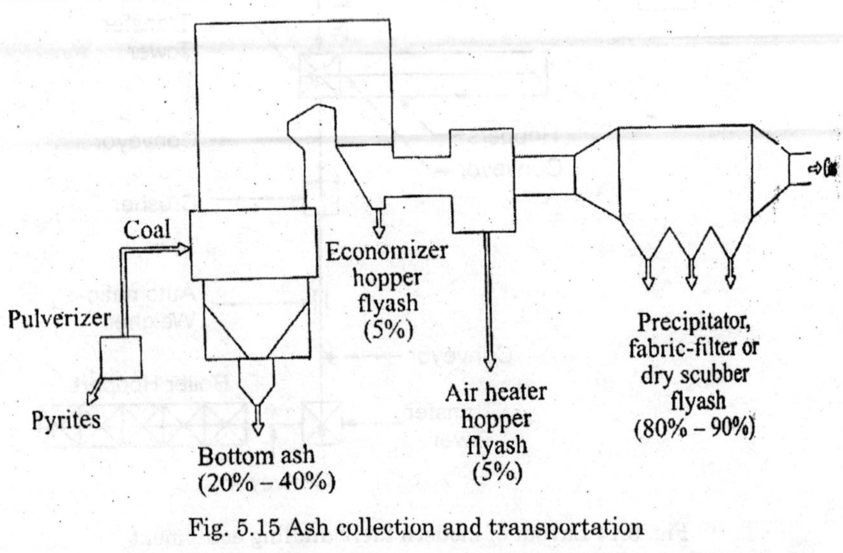

Fig. 5.15 Ash collection and transportation

Ash may be collected and transported from various points of the plants as shown in Fig. 5.15. A general outline of ash handling equipment is shown in Fig. 5.16.

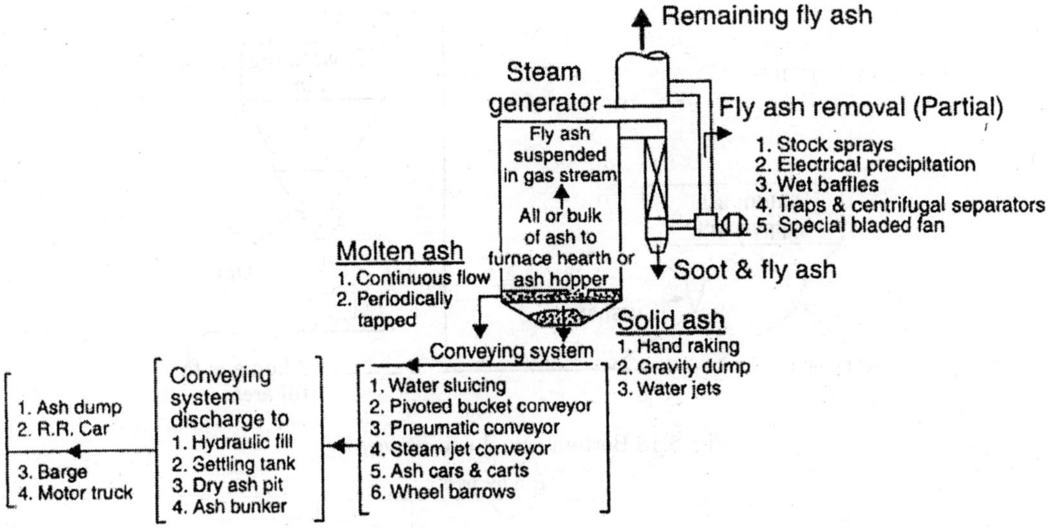

Fig. 5.16 General outline of ash handling equipment

Fig. 5.17 shows a general layout of ash handling and dust collection system which is self explanatory

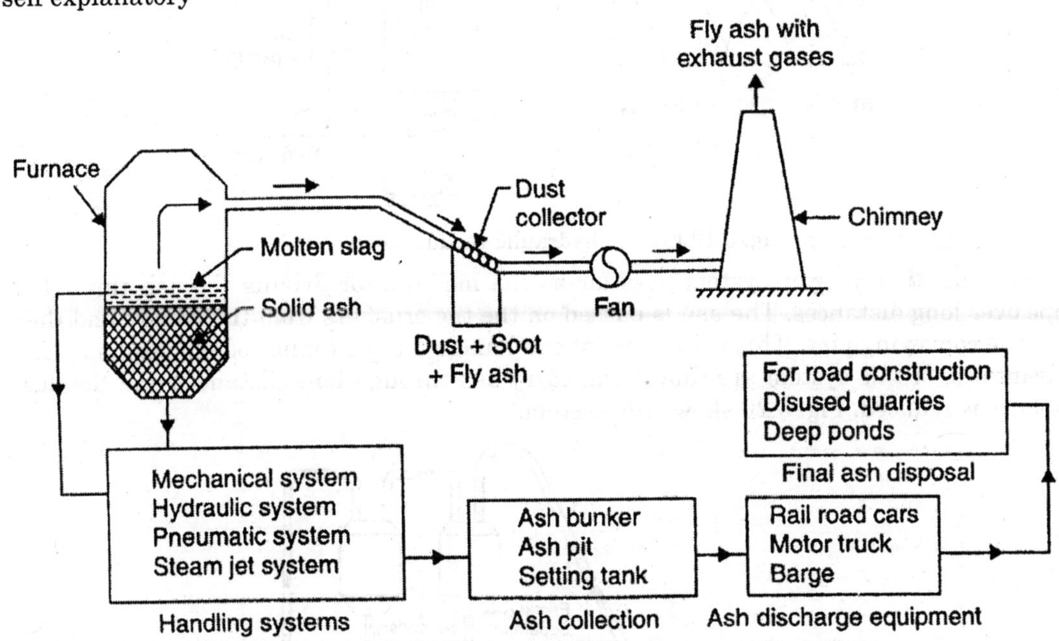

Fig. 5.17 Ash handling and dust collecting system.

The sluice conveyor system as shown in Fig. 5.18 is the most widely used for bottom ash handling, while the hydraulic vacuum conveyor as shown in Fig. 5.19 is the most frequently used for fly ash system.

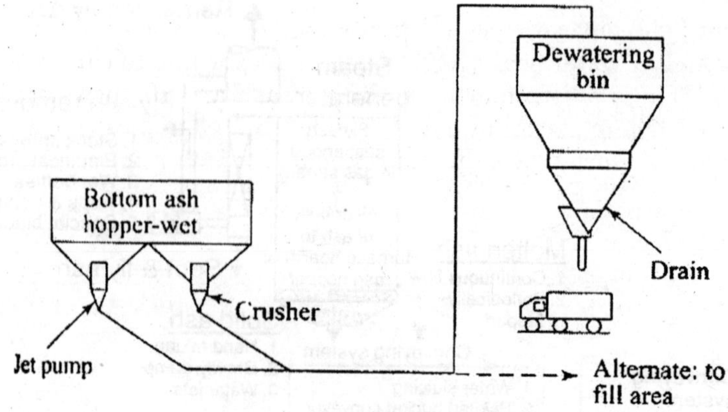

Fig. 5.18 Bottom ash sluice conveyor.

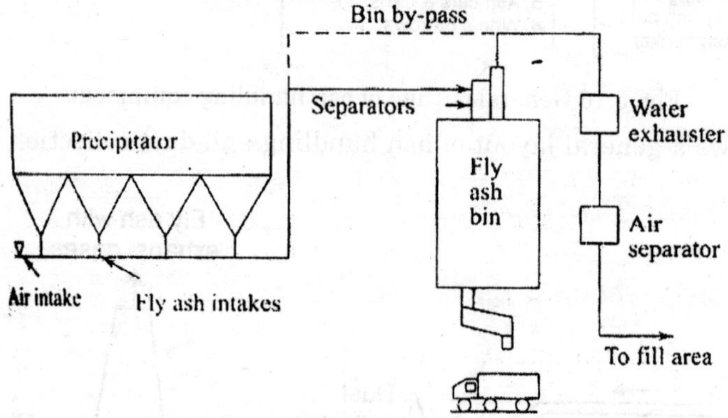

Fig. 5.19 Fly ash hydraulic vacuum conveyor

Pneumatic conveyor system uses air as the medium for driving the ash through a pipe over long distances. The ash is passed on the the crushers from the boilers and then into the conveying pipe. This arrangement can handle 5 to 30 tonnes of ash per hour. The advantages of this system are that it can carry ash through long distances and the dust mennce is reduced. Fig. 5.20 shows the system.

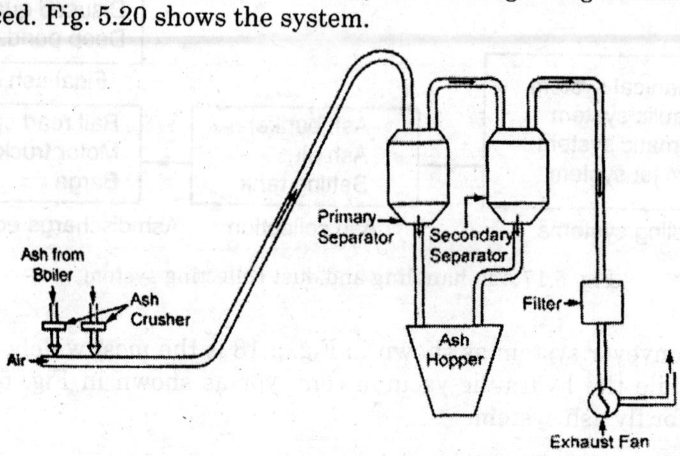

Fig. 5.20 Pneumatic conveying system

Fig. 5.21 shows belt conveyor system. Ash is hot when it comes out of the boilers. It is made to fall through a water seal over the belt conveyor and is carried to a dumping skite over the belt. This is a continuous handling of ash and the power consumption is low. With a speed of 300 mm per minute it can deliver upto 3 tonnes of ash per hour. Life of the belt under normal working conditions is about five years. A 20° rise in gradient is possible.

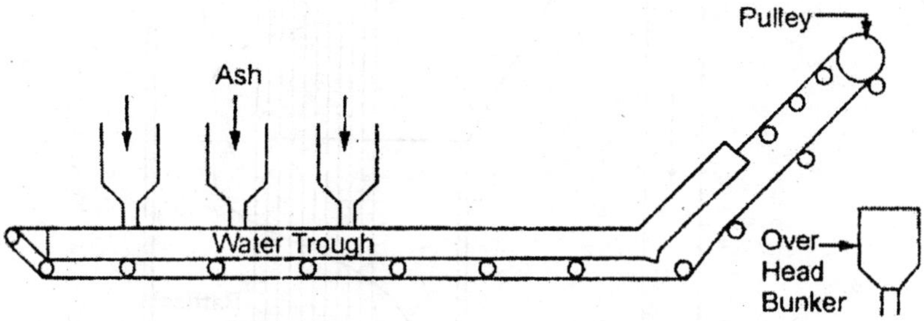

Fig. 5.21 Belt conveyor system.

5.7 AIR PREHEATER

Air preheaters are generally divided into two types, the recuperative and regenerative.

In recuperative air preheaters, the heat is directly transferred from the hot gasses to the air across the heat exchaning surface. They are commonly tubular although some plate types are still in use. Tubular units are essentilly counter flow shell and tube heat exchangers in which the hot gases flow inside the vertical straight tubes and the air flows outside as shown in Fig. 5.22 and air preheater arrangement shown in Fig. 5.23.

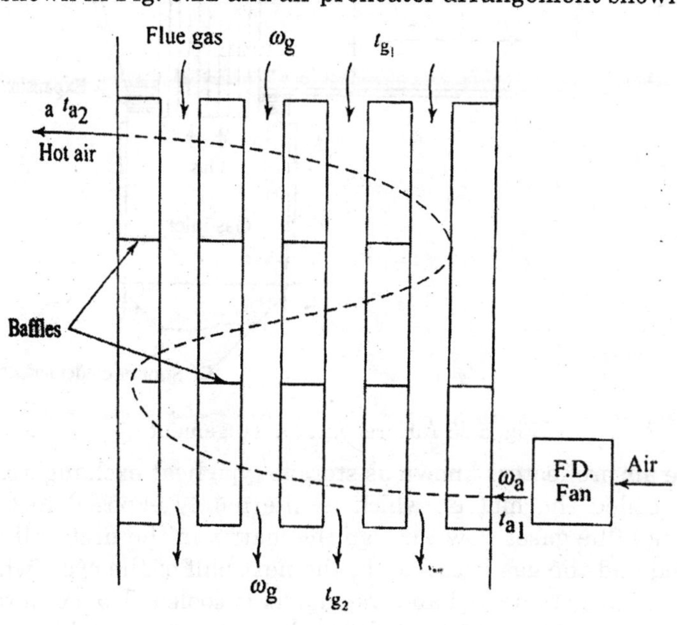

Fig. 5.22 Tubular air preheater

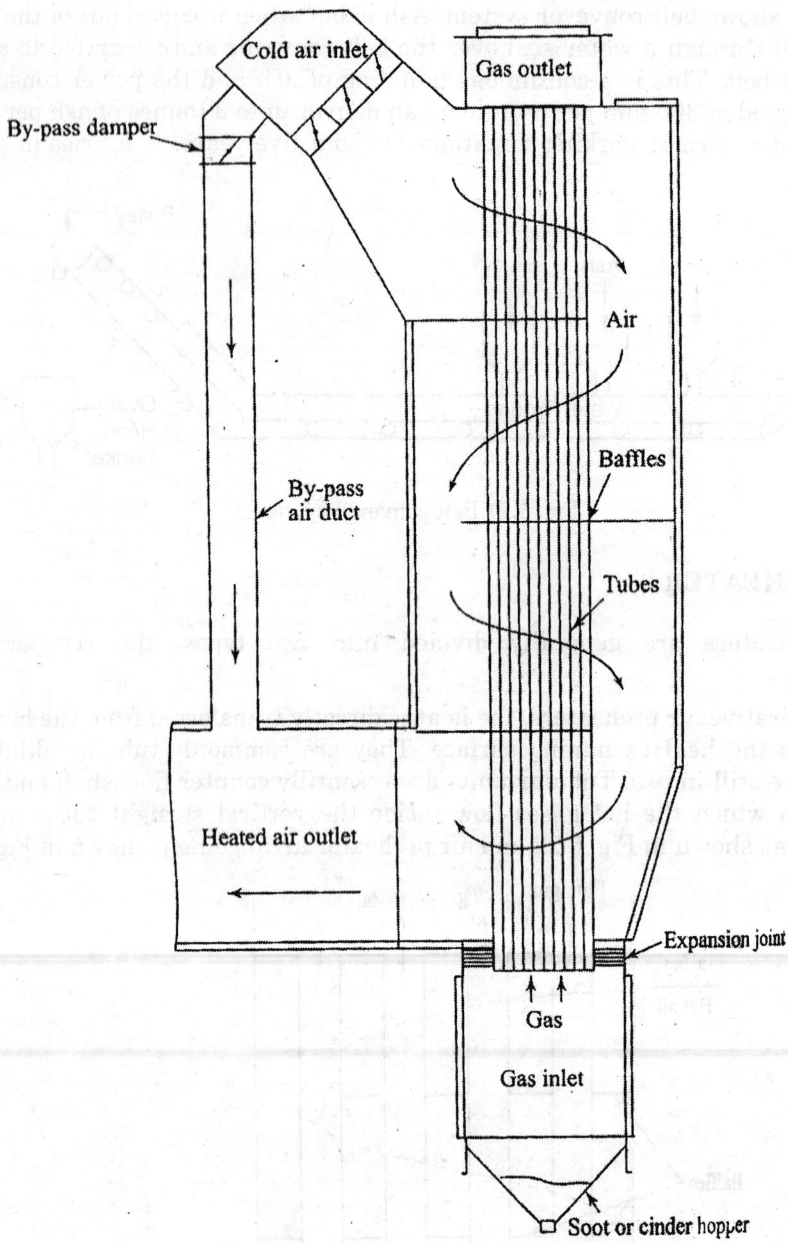

Fig. 5.23 Air preheater arrangement

Regenerative air preheaters known as storage type heat exchangers, have an energy storage medium, called the matrix, which is alternately exposed to the hot and cold fluids. When the hot flue gases flow through the matrix in the first half of the cycle, the matrix gets heated and the gas is cooled. In the next half of the cycle when the air flows through the matrix, air gets heated and the matrix is cooled. The cycle repeat itself. The heat transfer from gas to air is thus periodic, the amount depending on the number of heating - cooling cycles executed per second. The most common type is the rotary air preheater, known as Ljungstrom air preheater, which consists of a rotor driven by an electric motor through reduction gearing at a low speed with 2 to 4 rpm. The rotor is divided into a number of radial sectors, 12 to 24 in number, each sector being filled with steel sheets, either flat or corrugated as shown in Fig. 5.24. These sheets form the heat transfer surfaces. Two blank opposite sectors act as seals to prevent leakage of air to the gas path and vice versa. As the rotating sector enter the hot gas zone, they are progressively heated by the gas storing energy as sensible heat. When they enter the air zone, they progressively give by this heat to the air.

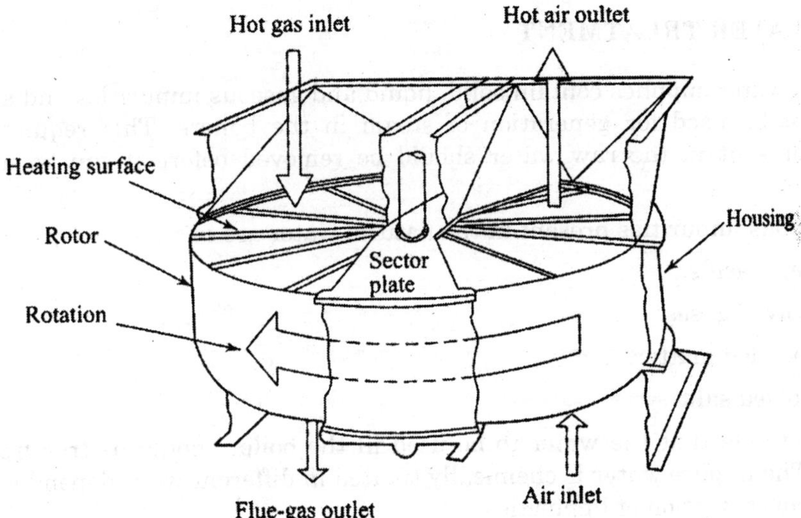

(a)

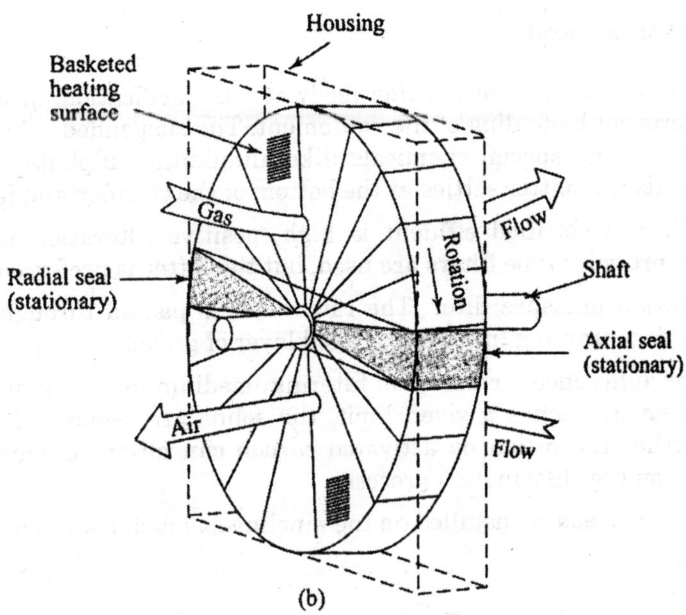

(b)

Fig. 5.24 (a) Regenerative air preheater (Ljungstorm air preheater)
(b) Another view of regenerative air preheater

Rotary air preheaters are compact heat exchangers with a large heat transfer surface being accommodated in small volume. They are relatively cheap and have good heat transfer characteristics. They are almost as popular as tubular air preheaters in power station practice. However, there are certain drawbacks:

(a) flow openings being plugged by fly ash,

(b) sealing of gas to air path and vice versa, and

(c) large pressure drops for both gas and air.

5.8 FEED WATER TREATMENT

Natural water supplies contain solid, liquid and gaseous impurities and as such this water cannot be used for generation of steam in the boilers. This requires that the impurities present in the raw water should be removed before it can be used in the boilers.

The various impurities present in the natural water are as :

(1) Mineral acids.

(2) Dissolved gases.

(3) Suspended matter.

(4) Dissolved salt.

It is desirable that the water to be used in the boiler should be free from various impurities. The impure water is chemically treated in different ways depending upon the nature and concentration of impurities.

5.8.1 Mechanical treatment

The first step of pretreatment of boiler feedwater is clarification, in which the water is chlorinated to prevent biofouling of the equipment. The suspended solids and turbidity are coagulated by adding special chemicals (like aluminium sulphate, $Al_2(SO_4)_3$) and agitated. The coagulated matter settles at the bottom of the clarifier and is removed.

If the turbidity of clarified effluent is high, positive filteration is needed. Both gravity filters and pressure type filters are used, but the latter is preferred.

Fig. 5.25 shows a pressure filter. The raw water is passed through alum pot and then through a tank containing fine sand, graded layer of gravel.

The pressure difference across the filtering medium is an indication of solid accumulation. When it reaches a given limit, the solids are removed from the bed by backwashing. Further filteration by activated carbon can absorb organics and remove residual chorine from the chlorination process.

These filters can be easily installed on the pipeline as much lesser space is required.

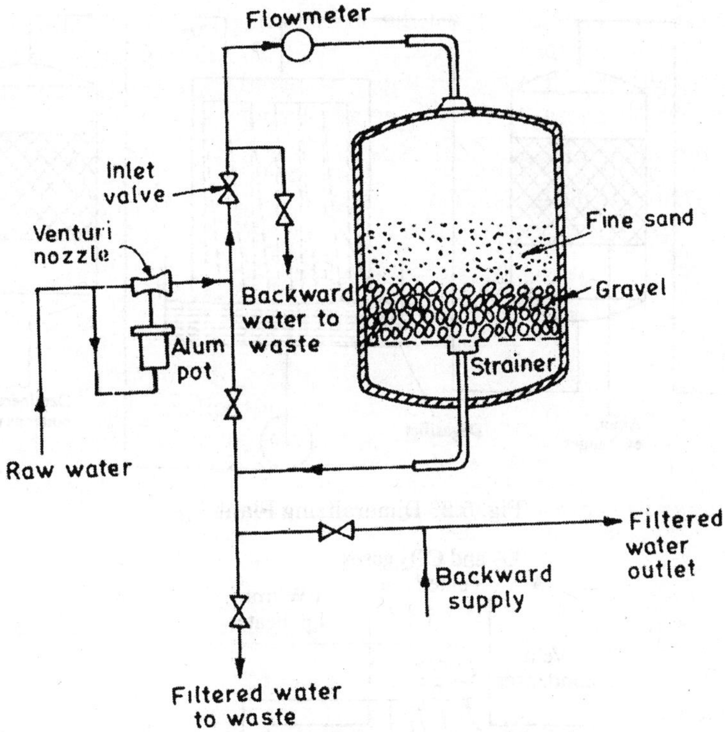

Fig. 5.25 Pressure type filter arrangement

5.8.2 Softening of water

Removal of hardness from water can be done by lime soda process, phosphate process, zeolite process and demineralization process.

The process of removing disolved solids in water by ion exchange is called demineralization. Two types of resins, cation and anion, are used. The cation resins is the hydrogen zeolite where the hydrogen ion is exchanged for the cation calcium, magnesium and sodium, and anion resin absorbs the anions chlorides, nitrates and sulphates, as discussed. Both ion-exchange processes are reversible, and the resins are restored to their original form by regeneration.

Fig. 5.26 shows a demineralizing plant consists of a cation exchanger, an anion exchanger, a degasifier and a silica adsorber.

1. Lime soda process.
2. Hot phospate softening.
3. Sodium zeolite softening.
4. Hydrogen zeolite softening.
5. Anion exchangers.
6. Condensate Polishing are not discussed here.

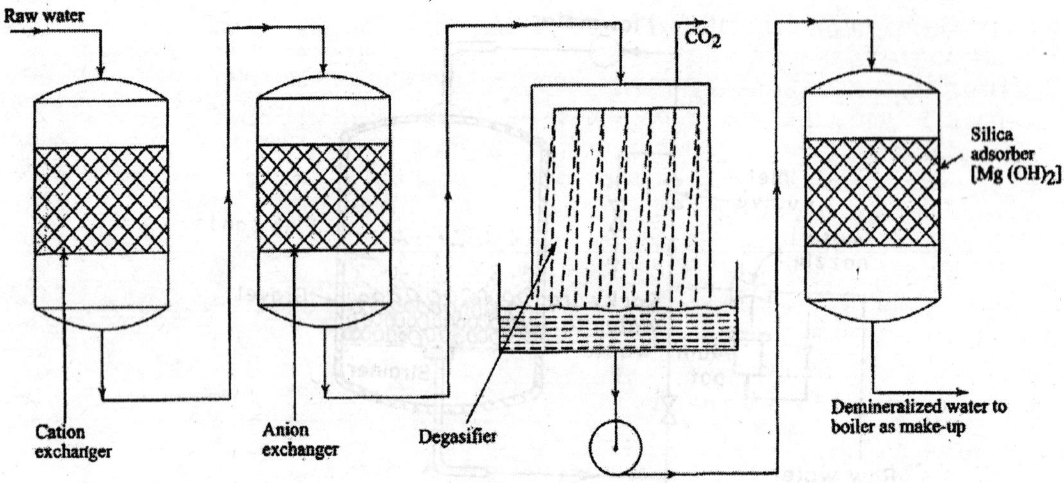

Fig. 5.26 Dimeralizing Plant

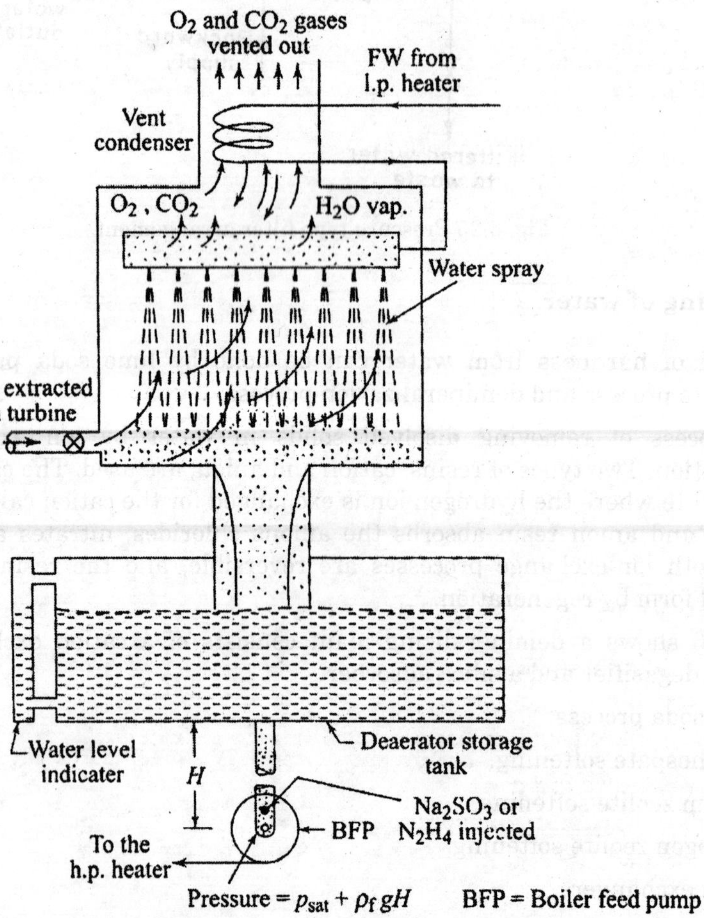

Fig. 5.27 Dearator with storage tank

5.8.3 Deaeration or degasification process

Deaeration or degasification is one of the most important steps in boiler water treatment. It depends on the decrease insolubility of dissolved gases, notably O_2 and CO_2. The solubility of these gases in water decreases with increase in temperature and becomes zero at the boiling or saturation temperature. These gases are removed in the deaerator, where feedwater is heated to the saturation temperature by the steam extracted from the turbine. Feedwater, after passing through a heat exchanger, called vent condenser, is sprayed from the top so as to expose large surface area, and the bled steam from the turbine is fed from the bottom. By contact the steam condensers and the feedwater is heated to the saturation temperature. Dissolved oxygen and carbon dioxide gases get released from the water and leave along with some vapour, which is condensed back in the vent condenser, and the gases are vented out as shown in Fig. 5.27.

5.9 FUEL COMBUSTION

Coal may be fed into the furnance for combustion in lump pieces or in powder form. Combustion of coal may occur in

(a) Fuel bed furnances (coarse particles).

(b) Pulverized coal furnance (fine particle).

(c) Cyclone furnance (crushed particles).

(d) Fluidized bed furnance (crushed small particles).

One of the chief factors in the economical working of a power plant is the efficient combustion of fuel. The two most commonly used methods for the burning of coal are stoker firing and pulverized fuel firing. The selection of one of the above methods depends upon the following factors.

1. Characteristics of coal available.
2. Capacity of boiler unit.
3. Station load factor.
4. Load fluctuations.
5. Reliability and efficiency of the various types of combustion equipment available.

5.9.1 Stoker firing

A stoker is power operated fuel feeding mechanism. Various advantages of stoker firing are as follows:

(1) Cheaper grade of fuel can be used.

(2) Higher efficiency attained.

(3) Greater flexibility of operation assured.

(4) Better furnance conditions maintained.

(5) Less smoke produced as compared to hand firing.

The main disadvantage of stoker firing is more cost of operation and repairing resulting from high furnance temperatures.

5.10 CLASSIFICATION OF STOKER FIRING

The working of stokers is based on the two principles.

1. Over feed stoker.
2. Underfeed stoker.

5.10.1 Over feed stoker

An over feed fuel bed section receives fresh coal on its top surface. It is characterized by the distinct zones from top to the bottom as shown in Fig. 5.28.

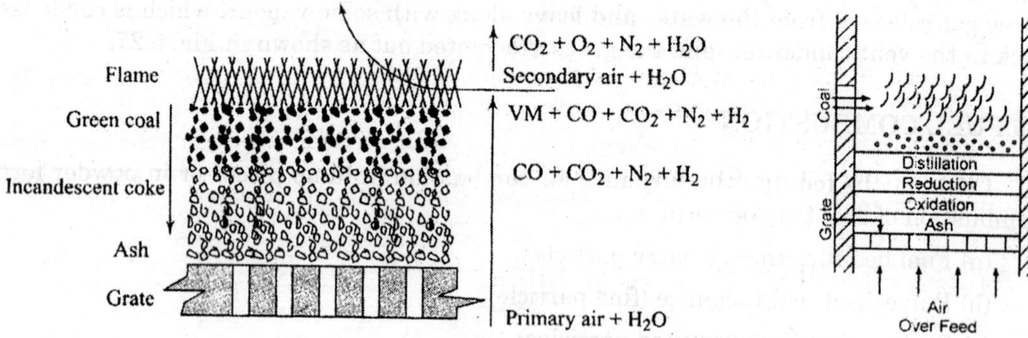

Fig. 5.28. Over feed stoker

1. A layer of fresh or green coal.
2. A layer of coal losing moisture - drying zone.
3. A coking layer of coal losing its volatile content distillation zone.
4. A layer of incandescent coke, where the fixed carbon is consumed combustion zone.
5. A layer of ash progressively getting cooler.

The primary air enters the grate from the bottom. The air while moving through the grate openings gets heated up and air while moving through the grate openings gets heated up and the grate is cooled. The hot air that moves through a layer of ash and picks up additional energy. The air then passes through a layer of in candescent coke where oxygen reacts with coke to form CO_2 and water vapours accompanying the air react with incandescent coke to form CO_2, CO and free H_2. The gases leaving the surface of fuel bed contain volatile matter of raw fuel and gases like CO_2, CO, H_2, N_2 and H_2O. Then additional air known as secondary air is supplied to burn the combustible gases. The combustion gases entering the boiler consist of N_2, CO_2, O_2 and H_2O and also CO if the combustion is not complete.

5.10.2 Underfeed Stoker

In underfeeding coal is fed from below the grate by a screw conveyor or ram as shown in Fig. 5.29.

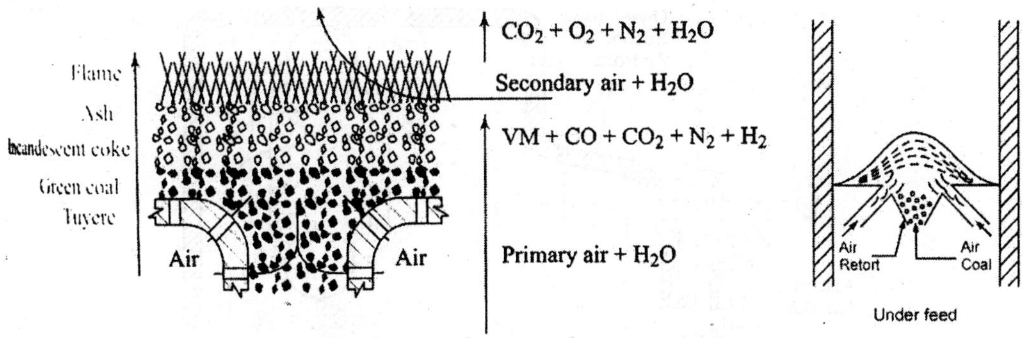

Fig. 5.29. Underfeed stoker principle

Primary air passing through holes in the tuyers diffuses through spaces in the raw or green coal picking up moisture and then through the distillation zone where volatile matter is added. When the gas stream passes hext through the incandescent coke region, the volatile matter breaks up and readily burns with secondary air fed at the top. In over feeding, burning the volatile matter will be somewhat cooler and so need longer time to ignite and burn.

5.11 TYPES OF MECHANICAL STOKERS

In small boilers, the grate is stationary and coal is fed manually by shovels. But for more uniform operating condition, higher burning rate and grate efficiency, moving grates or stokers are employed. Stokers may be of the following types

1. Travelling grate stoker.
2. Chain grate stoker.
3. Spreader stoker.
4. Vibrating grate stoker.
5. Underfeed stoker.

5.11.1 Chain grate stoker

Chain grate stoker and travelling grate stoker are similar in appearance but difference in the construction of grate.

Chain grate consists of flexible endless chain or crawlers of heavy machines and travelling grates consists of grate bars carried by steel bars.

Fig. 5.30 shows the chain grate stoker

The coal bed thickness is shown for all times by an index plate. This can be regulated either by adjusting the opening of fuel grate or by the speed control of the stoker driving motor.

The air is admitted from the underside of the grade which is divided into several compartments each connected to an air duct. The grate should be saved from being overheated. For this, coal should have sufficient ash content which will form a layer on the grate.

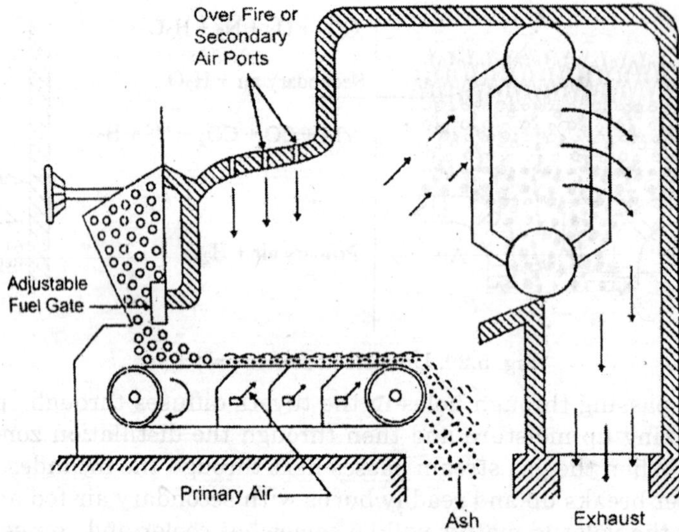

Fig. 5.30 Chain grate stoker

Since there is practically no aditation of the fuel bed, non coking coals are best suited for chain grate stokers.

5.11.2 Travelling grate stoker

The grate surface is made up of a series of CI bars joined together by links to form an endless belt running over two sets of sprocelet wheeels with a surface as wide as needed as shown in Fig. 5.31.

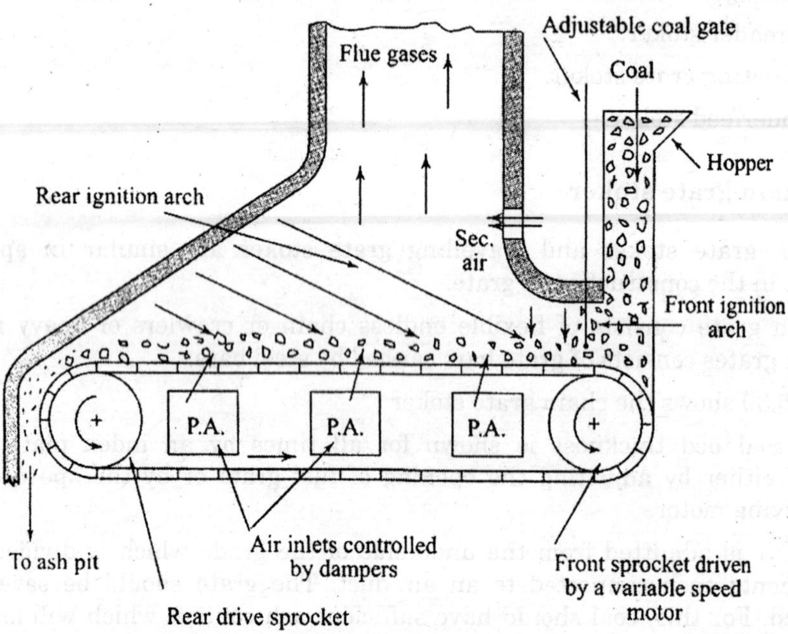

Fig. 5.31 Travelling grate stoker

These overfeed stokers are suited for industrial power plants having steady demand. The grate heat release rate should be limited to a maximum of 1340 kW/m^2 of active grate area. Coal having more than 20% ash is difficult to burn efficiently, incurring considerable unburn carbon loss.

5.11.3 Spreader stoker

Coal from the hopper is fed by rotating feeder, a drum fitted with short blader on its surface, to the spreader or distributor below, which projects the coal particles in a continuous stream on to the grate holding an ignited fuel bed as shown in Fig. 5.32. The finer particles burn in suspension and the coarser particles are consumed on the grate. The speed of the feeder varies directly with the steam output of the boiler.

Secondary or overfire air promote turbulence and completes combustion. High capacity boiler may have a travelling grate in addition to the spreader.

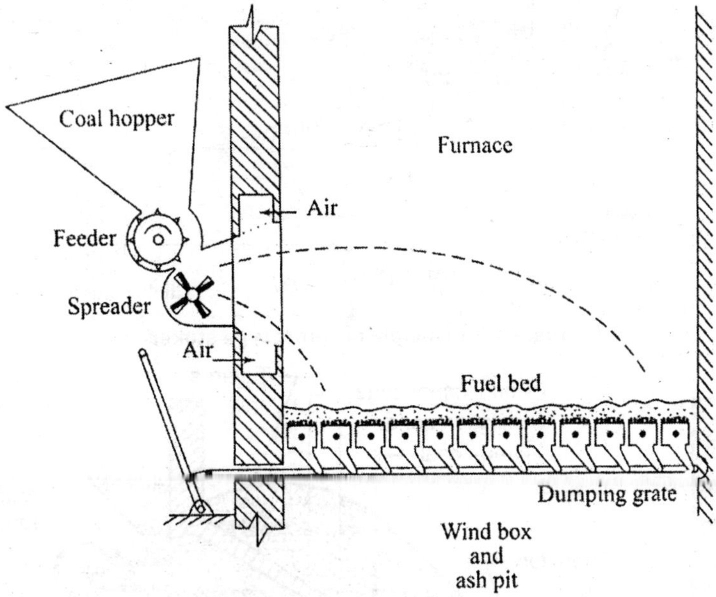

Fig. 5.32 Spreader stoker

5.11.4 Vibrating grate stoker

The stoker shakes the fuel bed intermittently, the frequency and amplitude of vibration depending on boiler load. The fuel bed is inclined so that the fuel moves towards the rear of the boiler by gravity with the progress of combustion and then fall into the ash pit. The grate is water cooled to prevent slagging.

5.11.5 Underfeed stokers

In underfeed stokers, the coal is fed into the furnace below the point for air admission. Coal from the hopper is pushed into the retort by means of reciprocating plunger. When coal gets heated, all the volatiles in it are distilled and when coal reaches the zone of the active combustion, it is in the form of coke and ash. The ash discharge plates are at the back of the furnace and by the time coal is pushed down on to these, all the combustion has been completed. Air is admitted into the furnace through holes in the sides of the retort. The coal is continuously agitated by the plunger and also by three pusher plates along the bottom of the retort. Due to this, the fuel bed remains prous and free from clinkers. Underfeed stokers are stable for non clinkering high voltatile coals having caking properties and low ash content.

Underfeed stokers, either single or multiple retort.

Fig. 5.33 shows the principle of retort type stoker and Fig. 5.34 shows multi retort under feed stokers

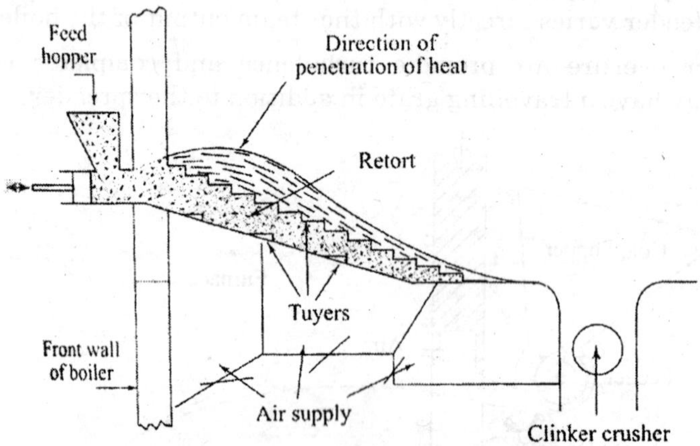

Fig. 5.33 Principle of retort type stoker

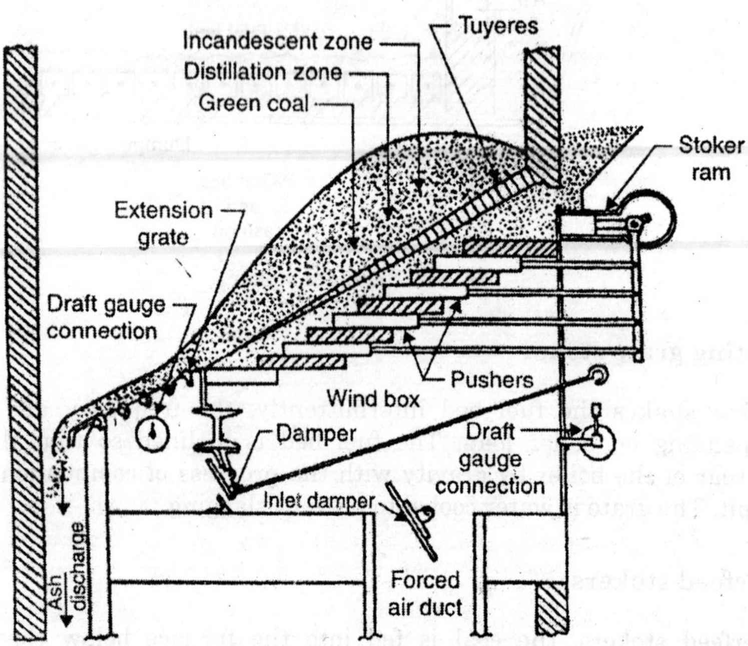

Fig. 5.34 Multi-retort underfeed stokers

5.12 PULVERIZED COAL FIRING SYSTEM

Solid fuels can be used in powdered state eiter alone or mixed with crude oil and burnt like gas or oil. Coal is reduced to a fine powder in grinding mills or pulversisers and then projected into the combustion chambers by means of a current of hot air. A further volume of preheated air to make up the necessary amount for combustion is blown in separately and the resulting turbulence in the high pressure combustion chamber helps in through combustion of the fuel.

The amount of air which is used to dry the coal and convey the powdered fuel to the furnace is known as "Primary Air" and the amount of air which is blown in separately to complete the combustion is known as "Secondary Air". Primary air and Secondary air system is shown in Fig. 5.35.

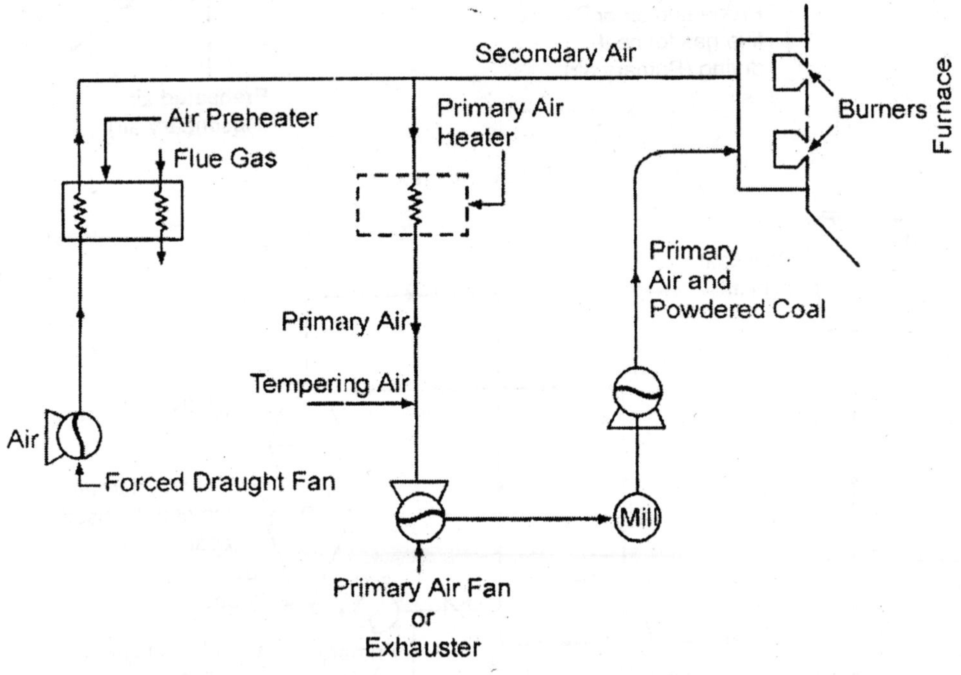

Fig. 5.35. Primary and secondary air system.

5.12.1 Unit and central system

There are two systems of preparation of coal into a fine powder. One is "Unit System" and the other is "Bin System" or "Storage system" or "Central system".

In Unit system, a separate pulverising unit is provided for each furnace and the fuel is fired directly into the furnace without being stored. Whereas in the central system, the fuel is ground in a central grinding plant and stored in bins or bunkers. From the bunkers, it is distributed to various burners through separate feeders in accordance with load demands. Fig. 5.36 shows a line diagram of Unit system and Fig. 5.37 shows a line diagram of central system.

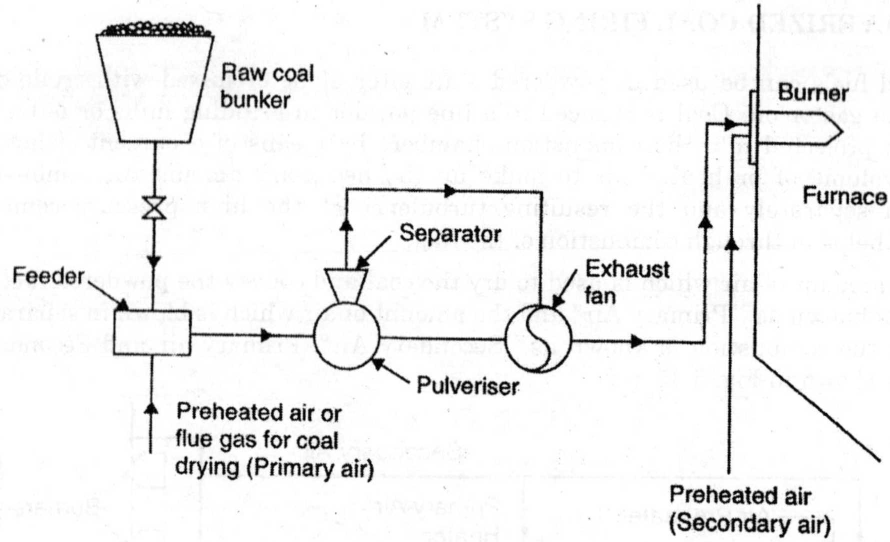

Fig. 5.36 Unit system

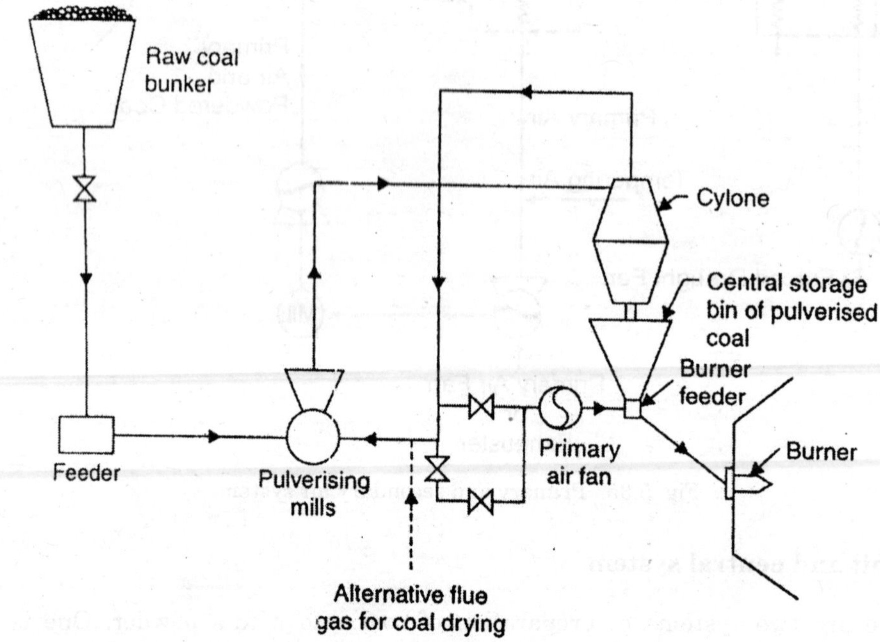

Fig. 5.37. Central system

5.12.2 Pulverizers

There are three stages in the pulverizing process of cola:

(a) Feeding.

(b) Drying.

(c) Grinding.

The feeding system controls the fuel feed rate according to the boiler demand and the requirement air rate (primary air) for drying and then transporting the pulverized fuel and primary the pulverized fuel and primary air stream to the burner. Grinding is performed by impact, attrition, crushing or combustion of all these is shown in Fig. 5.38.

Pulverizers, commonly used, are classified by speed.

(a) Low speed (below 75 rpm), ball tube mill

(b) Medium speed (75 to 225 rpm), the ball-and-race mill and bowl mill and (c) high speed (about 225 rpm), the impact or hammer mill.

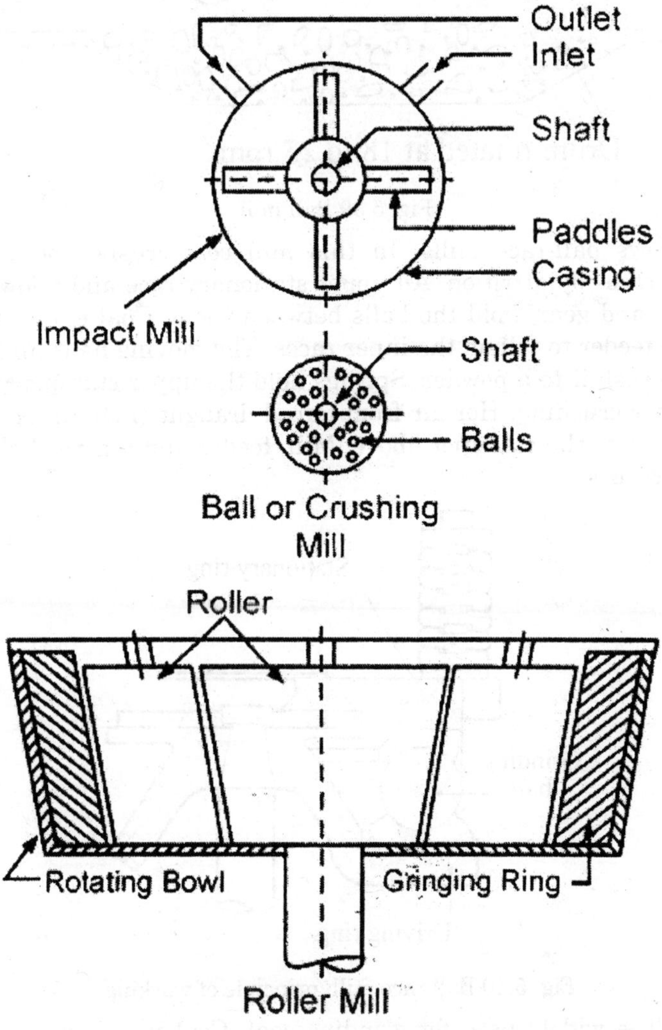

Fig. 5.38 Pulverizers principle

The ball tube mill is a hollow horizontal cylinder with conical ends are wear resistant liners revolving slowly at about 20 rpm with 20 to 35% of its volume being filled with forged steel balls or mixed size 30 to 60 mm as shown in Fig. 5.39.

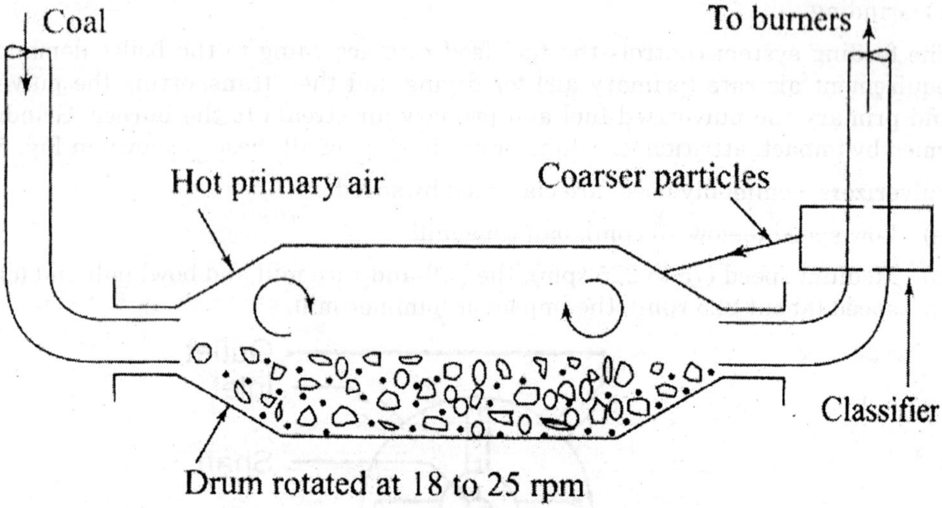

Fig. 5.39 Ball mill

Fig. 5.40 shows ball-race mills. In this mill coal crushes between two moving surfaces ball and races by attrition. An upper stationary race and a lower rotating race, driven by a worm and gear, hold the balls between them. Coal enters the mill through the rotating table feeder to fall on the inner races. The moving balls and races catch coal between them to crush it to a powder. Springs hold the upper stationary race and adjust the force needed for crushing. Hor air from forced draught (FD) fan enters the mill and carries the coal dust to the classifier above. Mill, feeder and fan need about 17 kWh per tonnes of coal pulverized.

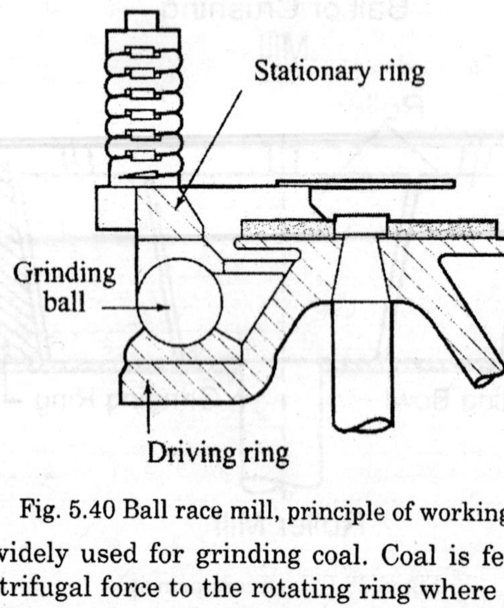

Fig. 5.40 Ball race mill, principle of working

The bowl mill is widely used for grinding coal. Coal is fed to a rotating bowl and thrown outward by centrifugal force to the rotating ring where it is pulverized by rolling under spring-loaded rollers as shown in Fig. 5.41. Air sweeps by the bowl and removes the product to the classifier, from which large particles are returned for further grinding. This is an intermediate speed pulverizer (vertical shaft speed about 65 to 100 rpm) which, with its quiet, vibrationsless operation, combines satisfactorily the desirable characteristics of low power consumption and low maintenance cost.

High speed pulverizers use hammer beaters that revolve in a chamber with weat resistant liners. They are mostly used with low rank coals with high moisture content and often use flue gas for drying.

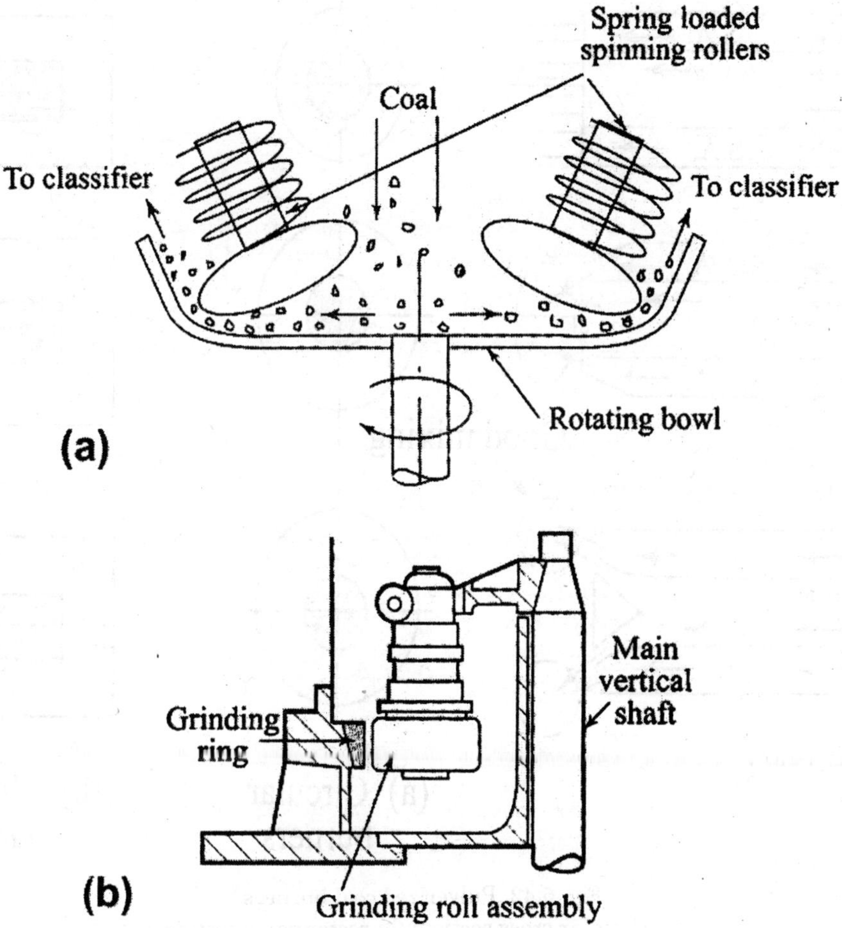

Fig. 5.41. (a) Bowl mill (b) Roller type ring-rong mill

5.12.3 Pulverized coal burners

The main difference between the various burners lies in the rapidity of air-coal mixing i.e turbulence. Pulverzed coal burners can be divided into two main groups: circular and slot type burners, as shown in Fig. 5.42. The simplest circular burner consists of a cylindrical tube through which the pulverized coal and primary air mixture is bloun into the furnance, the secondary air being delivered through a separate tube. The simplest slot type burner differes from the circular burner only in the shape of its cross section.

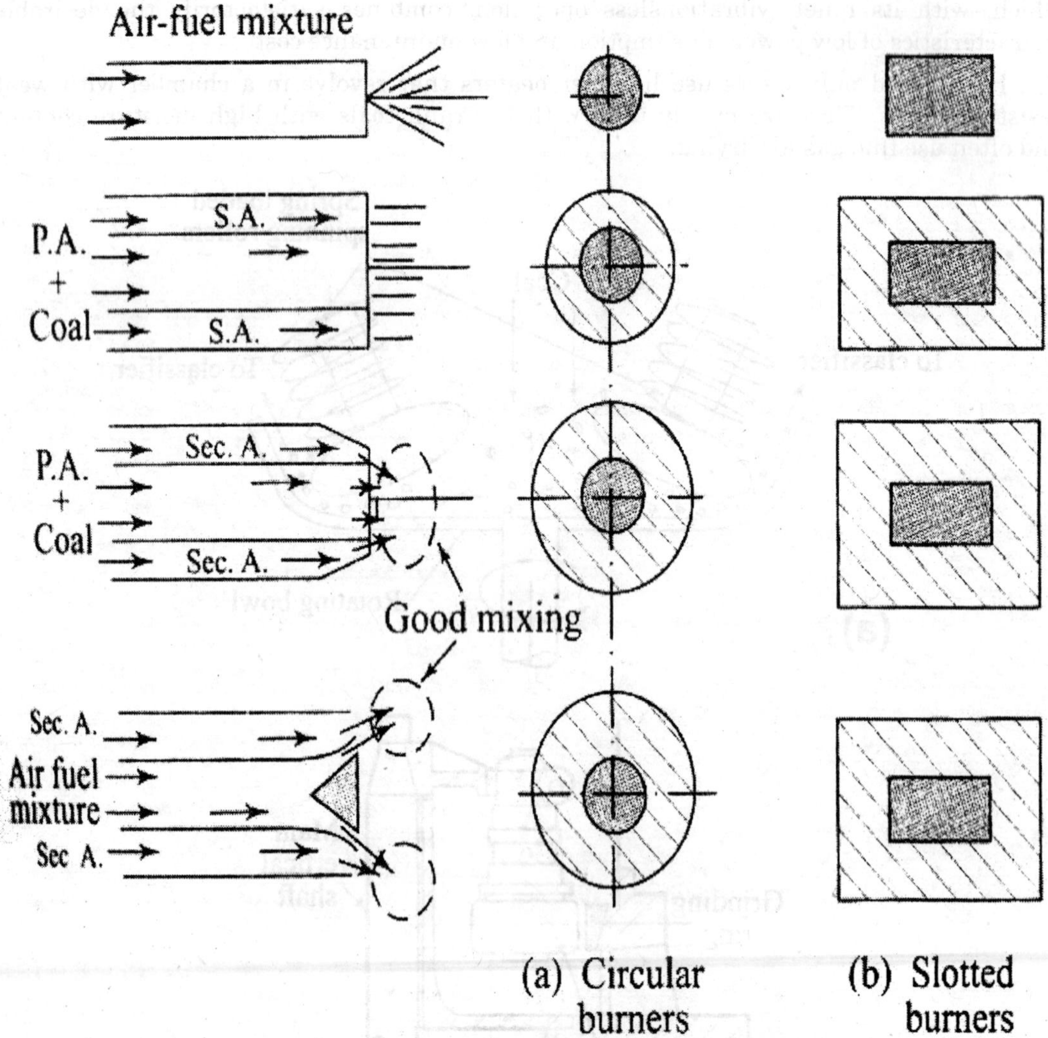

Fig. 5.42. Pulverized coal burners
(a) Circular cross section; (b) rectangular section

Burners are arranged on the furnance walls so as to ensure almost complete combustion of fuel in the flame core itself. Burners are optimally arranged depending on their type and characteristics.

1. **Straight flow burners**: In this type complete combustion can be achieved by huge mixing of fuel from different burners upon their collision in the furnace space. Typical arrangements of these arrangements are shown in Fig. 5.43.

Fig. 5.44 shows the flow patterns of flame in dry bottom furnaces.

2. **Cyclone furnace:** It is widely used to burn poorer grades of coal having high ash and moisture content. Biofuels like rice husks can also be successfully burned in cyclone furnances for steam generation.

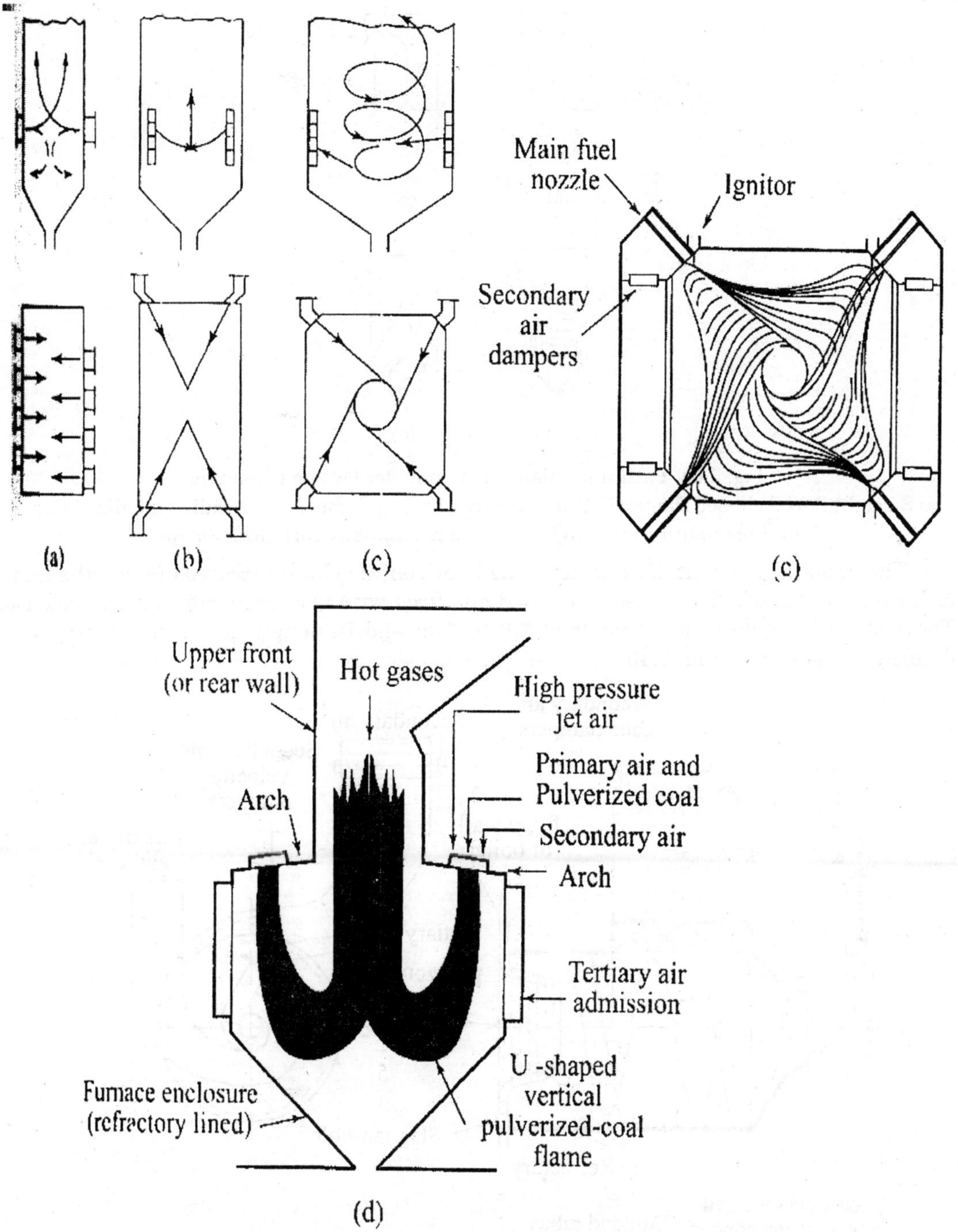

Fig. 5.43. Arrangement of straigh flow burners on furnance walls
(a) opposed displaced firing (b) corner firing with encountering jets
(c) corner firing with tangential jets (d) Vertical firing

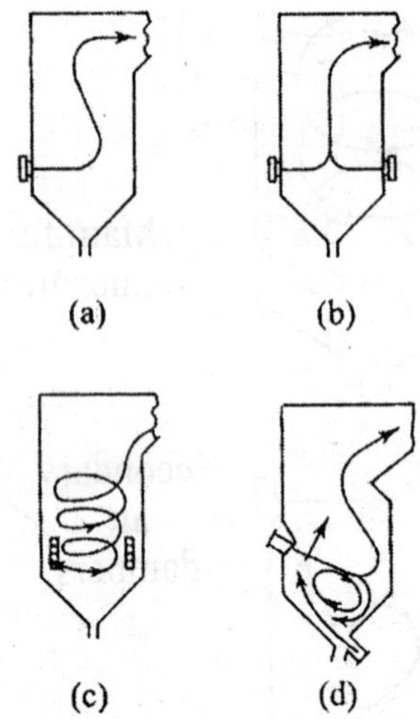

Fig. 5.44. Patterns of flame motion in dry bottom furnances
(a) Straight flow S shaped flame; (b) Opposite straight flow flame; (c) Vertically turbulized flame; (d) Combination of straight flow and horizontally turbulized flame.

The cyclone is essentially a water cooled horizontal cylinder located outside the main boiler furnace, in which crushed coal is fed and fired with very high rates of heat release. The cyclone is made to a diameter of 1.8 to 4 m and its length is 1.2 to 1.3 times its diameter on shown in Fig. 5.45.

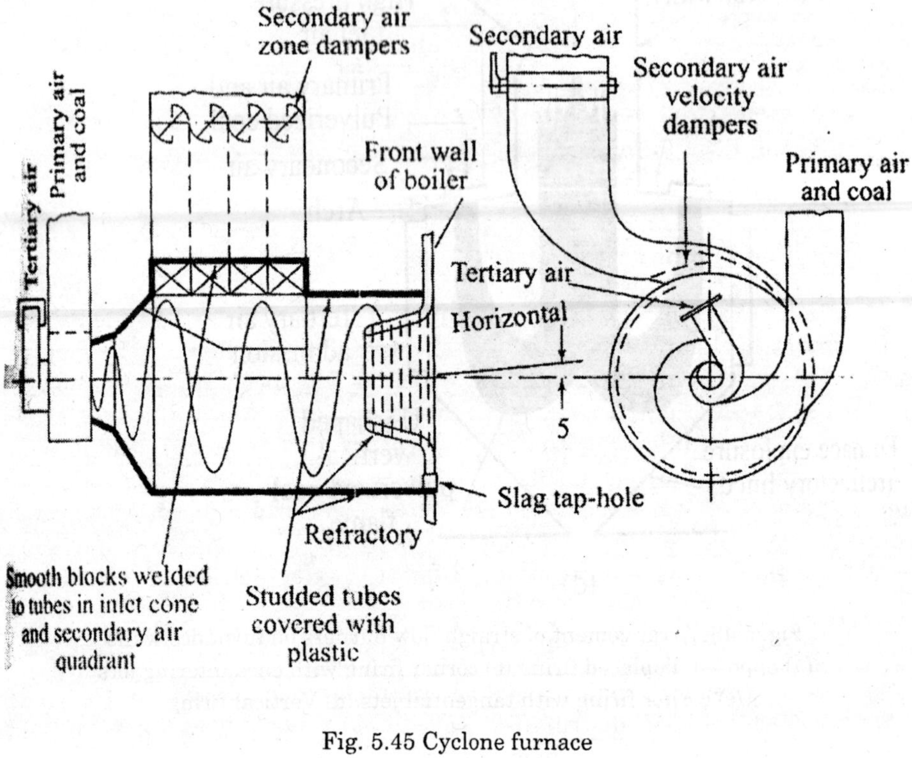

Fig. 5.45 Cyclone furnace

3. Turbulent burners: These burner give a shorter and wider flame than do the straight flow type. The primary and secondary air flows are intensively intermixed due to the energy of turbulent motion, which secures a thorough burning off of fuel in the flame cope.

The arrangement of turbulent burners are shown in Fig. 5.33.

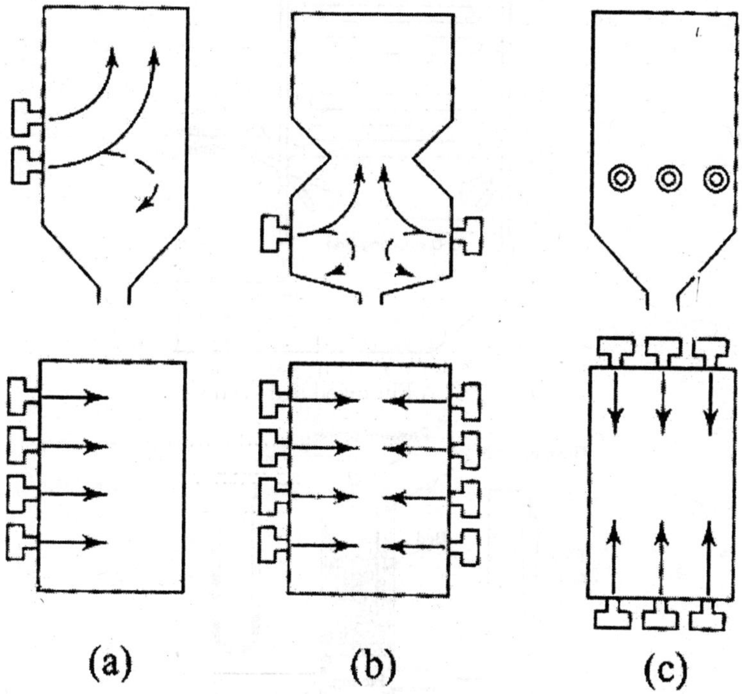

Fig. 5.46 Arrangement of turbulent burners on furnace walls.
(a) Front firing; (b) Opposed firing; (c) Opposed firing on side walls

5.12.4 Fluidized bed combustion

When air is passed through a fixed or packed bed of particles, air simply percolates through the interstitial gaps between the particles. As the air flow rate through the bed is steadily increased, a point is eventually reached at which the pressure drop across the bed becomes equal to the particles per unit cross sectional area of the bed. This critical velocity is called the minimum fluidization velocity at which the bed is said to be incipiently fluidized. As the air velocity is increased further, the particles are buoyed up and imparted a violently turbulent fluidlike motion, with the drag forces exerted by the fluid on the particles exceeding their weight. There is a high degree of particle mixing and equilbrium between gas and particles is rapdily established. This is called a fluidized bed.

Fig. 5.47 shows a fluidized bed combustion system.

In Fig. 5.48 (a), air supplied by a centrifugal blower is passed through a perforated or porous plate, called the distributor, and then a bed of particles of wide size distribution. A few distributors are shown in Fig. 5.48 (b).

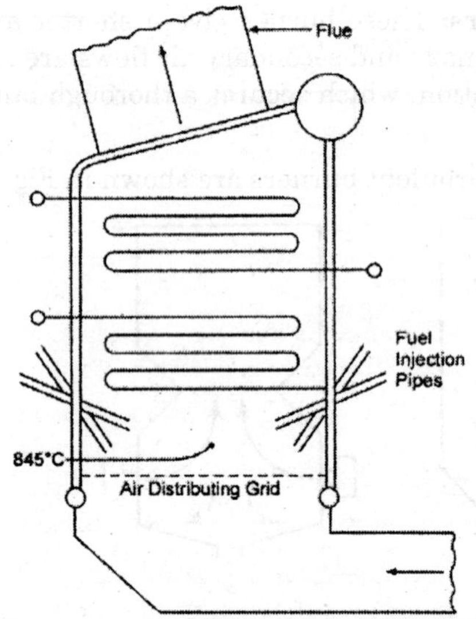

Fig. 5.47 A Fluidized bed system

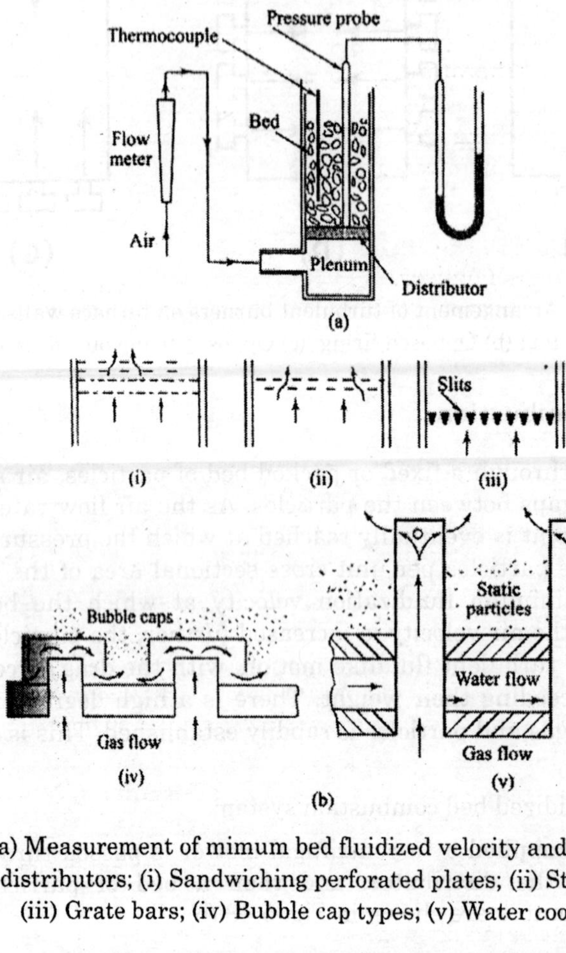

Fig. 5.48 (a) Measurement of mimum bed fluidized velocity and U+ of fluidized bed; (b) Few types of distributors; (i) Sandwiching perforated plates; (ii) Staggered perforated plates; (iii) Grate bars; (iv) Bubble cap types; (v) Water cooled type

5.13 TYPES OF BOILERS

The water tube boilers are best suited for high pressure, high capacity steam generators. All water tube generators may be classified as either natural-circulation or forced circulation boilers. Most of the older steam generators are natural circulation systems.

To increase the rate of heat transmission in boilers it is simpler to employ high water velocities rather than high gas velocities, because a smaller quantity of fluid is dealt with, and a considerable increase in pressure is more easily produced with water than with gas.

Moreover, as the pressure of steam increases its density also increases whilst the corresponding increase in temperature causes the density of the water to decrease. This double change reduces the thermo-syphonic head upon which natural circulation depends, so that a point is reached where forced circulation becomes imperative.

A number of modern boilers used forced circulation rather than natural circulation operation of such steam generators is basically similar to that of a natural convection unit. Such units are particularly useful at very high pressures that is above 160 bar, where the reduced density difference between liquid and vapour limits the natural circulation heat available.

High pressure steam generators currently operating in electric power stations. Steam generating capabilities for these boilers range from about 135000 kg/h to an undermined maximum exceeding 3.2×10^6 kg/h. The operating steam pressure is subscritical, (115 bar to 165 bar). Steam superheat and reheat temperatures as usually 538°C.

The advent of the water cooled furnace walls, called water walls, eventually led to the integration of furnace, economiser, boiler, superheater, reheater and air preheater into the modern steam generator. Water cooling is also used for superheater and economiser compartment walls and various other components, such as screens, dividing walls, etc. The use of a large number of feedwater heaters (upto seven or eight) means a smaller economiser, and the high pressure means a smaller boiler surface because the latent heat of vaporization decreases rapidly with pressure. Thus a modern surface high pressure steam generator requires more superheating and reheating surface and less boiler surface than older units.

Three design concepts of water tube boilers are illustrated in Fig. 5.49.

Natural circulation is used upto steam pressures of approximately 180 bar, with separation of the steam from the water taking place in the boiler drum. [Fig. 5.49(a)]

Boilers with forced circulation by a special pump, originally known as La Mont boilers [Fig. 5.49 (b)]. These type of boilers are suitable for steam pressures up to 200 bar. Provilers operating at subcritical pressures (Less than 221.2 bar) which rely on a drum and recirculation, either natural or forced, are commonly known as drum boilers. Circulation in the boiler is often controlled by orifizers. usually located at the bottom of the tubes. They ensure even distribution of flow through the tubes in the waterwalls.

Boilers can be designed to operate at pressures above the critical pressure (221.2 bar). No drum is necessary in such boilers since no separation of the water and steam occurs; the two being at the same density. Thus there is no circulation. The water enters the bottom of the tubes and is completely transformed to steam by the time it reaches the top, passing through the tubes only once. For this reason it is known as a once-through boiler, often referred to as drumless boiler. It is illustrated in Fig. 5.49 (c) and 5.49 (d).

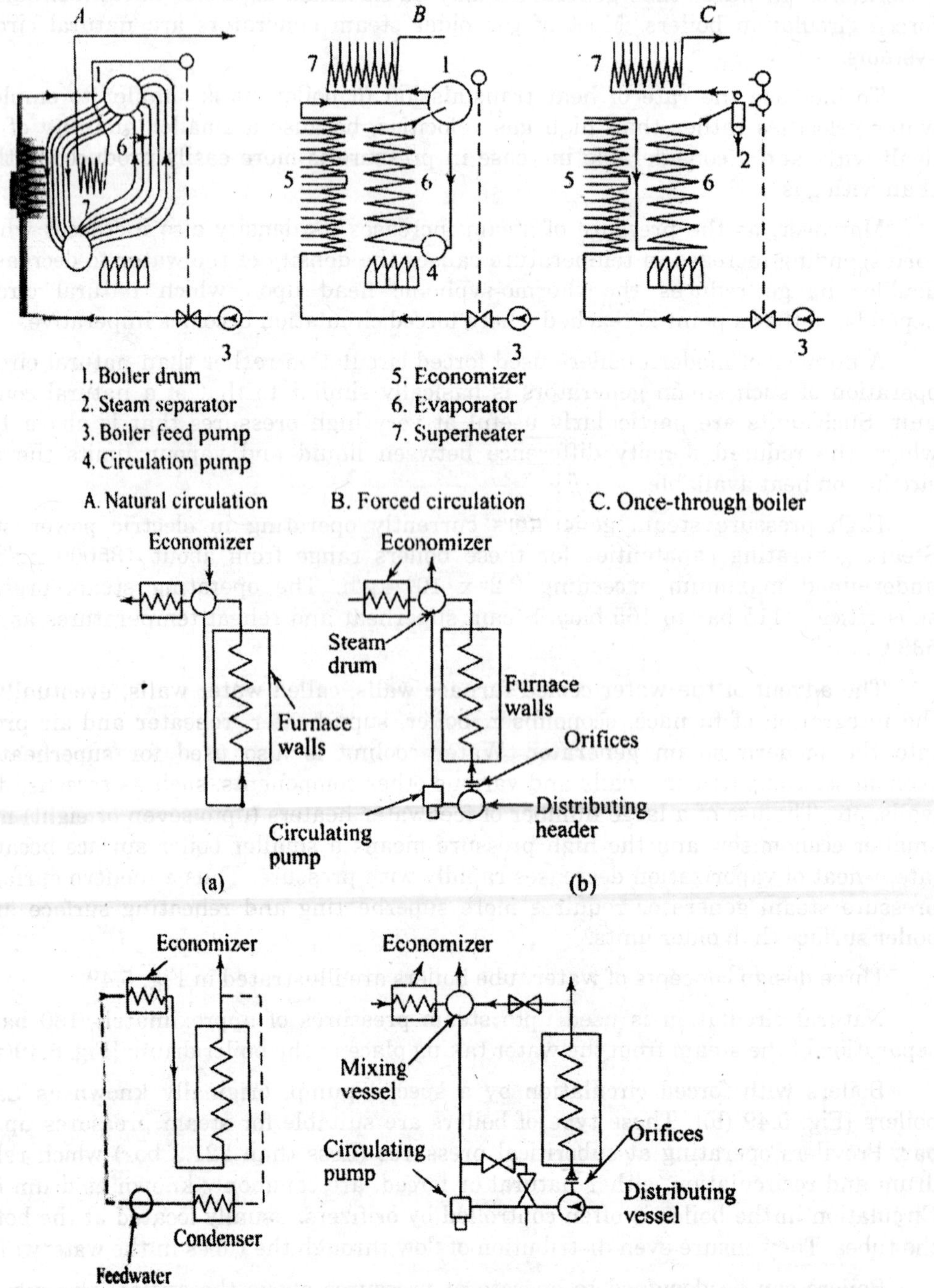

Fig. 5.49. Types of boiler circulation

5.14 MODERN DEVELOPMENTS IN STEAM BOILERS

A number of modern boilers use forced circulation rather than natural circulation.

The La Mont boiler, shown in Fig. 5.50 closely resembles the natural circulation boiler because it has a steam drum. Since the water steam mixture is separated in the drum, the system must operate at a pressure below the critical pressure.

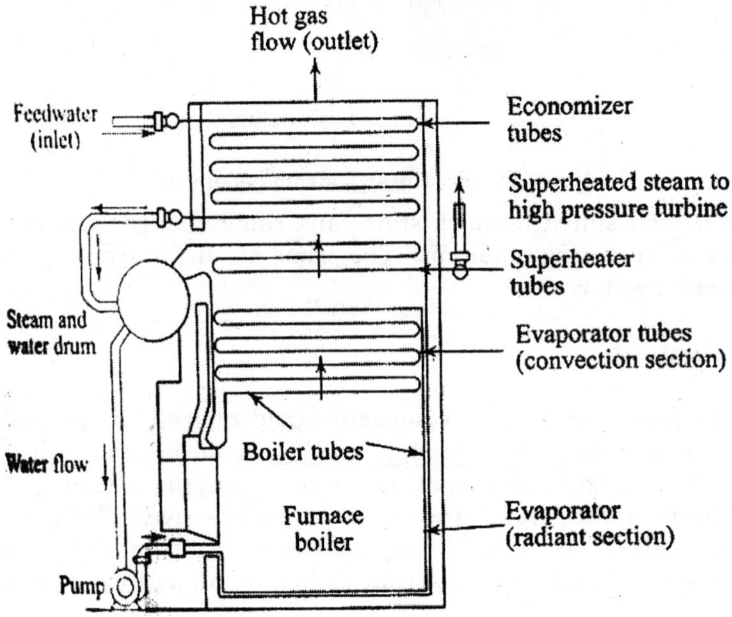

Fig. 5.50 La-Mont boiler

The once through principle of steam generation in boilers assumes that the quality of water fed to the heated element (tube, loop, row of loops connected in parallel) with the aid of a pump is equal to the amount of steam (vapour) formed in this element. The simplest once through boiler is a coil (loop) in which water is delivered into one end and superheated steam emerges from the other. In order to construct high power once through boilers of a reasonable hydraulic resistance to fluid flow the heating surfaces of such boilers are built up a large number (several dozens) of parallel loops. A schematic diagram of a once through boiler (Fig. 5.51) shows that the zone of the begining and end of vaporization may shift, depending on operating conditions, i.e. load, kind and quality of fuel fired etc. Contrary to natural circulation boilers, once through boiler units comprise no heavy and costly elements, such as the drum, system of non heated down comer tubes and headers.

There are many kinds of forced circulation boilers, depending upon the circulation paths in the evaporator. The most widely used forced circulation boiler system in the USA is the universal pressure or Benson boiler. The most widely used in the USSR is the Ramsin boiler. Benson and Ramsin boilers are known as once through steam generator.

The once through boiler is the only type suited to supercritical pressure operation (above 221.2 bar).

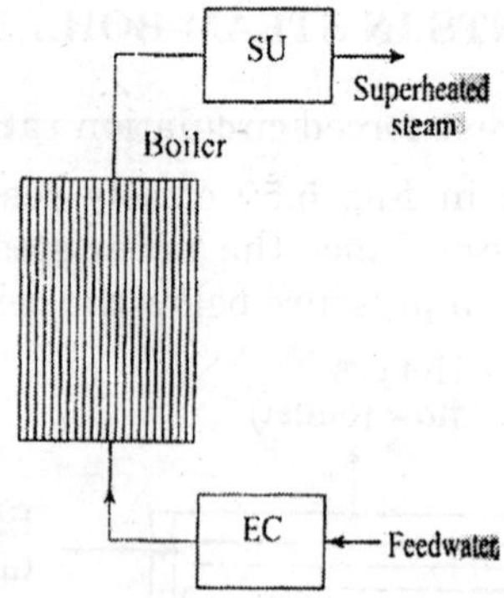

Fig. 5.51 Once through steam generator

The design is well suited to quick starts and rapid changes and provides positive circulation under all operating conditions. The prime disadvantage is the requirement of extremely pure boiler water.

5.15 COOLING TOWER

A generating station of the given capacity would typically be located near a large body of water or a river and "once through" cooling would have been used in the part. Such cooling involves taking water from, say, a river, passing it through the condenser tubes, and discharging the heated water back into the river. With increased power generation and stringent ecological and environmental standards imposed by the government, through cooling is seldom permitted in new, high capacity power stations.

In such cases, the waste heat must be rejected to the atmosphere, with cooling water recirculating and serving as a transport medium for heat between the source and sink. One way of achieving this is the use of cooling towers.

In the cooling tower, the water is in direct contact with the air. The water temperature may be theoretically cooled to the wet bulb temperature of the air. Since the cooling tower utilizes the phenomenon of evaporative cooling, much more heat can be transferred to the air with a cooling tower than with a simple heat exchanger.

A schematic diagrams of three types of cooling towers are shown in Fig. 5.52.

5.15.1 Induced draught cooling tower (Forced draught cooling tower)

An induced draught counter flow cooling tower is shown schematically in Fig. 5.53. Air is drawn into the tower from the bottom and leaves at the top. Warm water from the condenser is pumped to the top of the tower and is sprayed into this air stream. The purpose of spraying is to expose a large surface area of water to the air. As the water droplets fall under the influence of gravity, a small fraction of water (usually a few percent) evaporates and cools the remaining water. The temperature and the moisture content of the air increase during this process. The cooled water collects at the bottom of the tower and is pumped back to the condenser to pick up additional waste heat. Makeup water must be added to the cycle to replace the water lost by evaporation and air draught. To minimize water carried away by the air, drift eliminators are installed in the cooling towers above the spray section. The packing of the tower is usually formed from wooden slats. A modern design of tower employs a plastic impregnated cellulose material which has high water absorption qualities and a long working life.

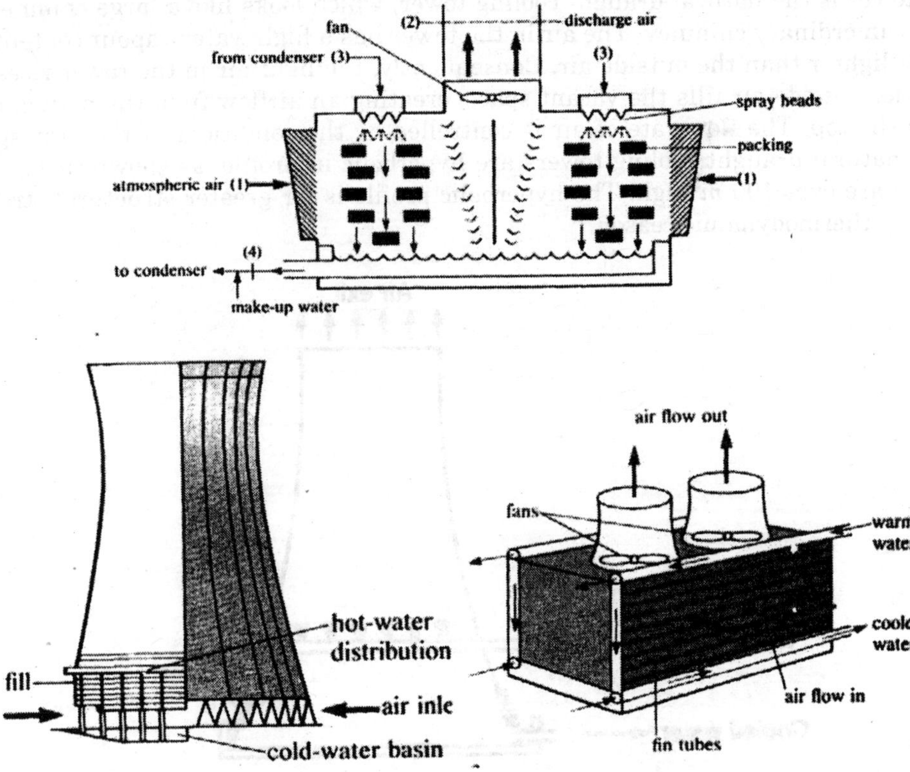

Fig. 5.52. Schematic diagrams of three types of cooling towers:
(a) An induced draft evaporative cooling tower; (b) A natural draft evaporative cooling tower; (c) A dry cooling tower

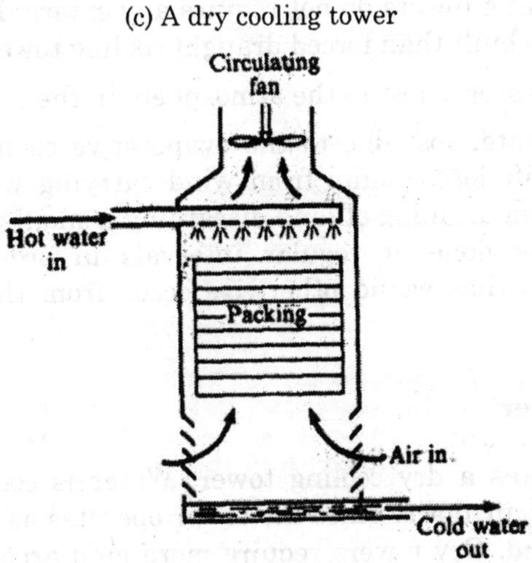

Fig. 5.53. Induced draught cooling tower

5.15.2 Natural draught cooling towers

The air circulation in the cooling tower is provided by a fan and therefore it is classified as a forced draught cooling tower (described above). Another popular type of cooling tower is the natural draught cooling tower, which looks like a large chimney and works as an ordinary chimney. The air in the tower has a high water vapour content, and thus it is lighter than the outside air. Consequently, the light air in the tower rises, and the heavier outside air fills the vacant space, creating an airflow from the bottom of the tower to the top. The flow rate of air is controlled by the conditions of the atmospheric air. The natural draught cooling towers are hyperbolic in profile, as shown in Fig. 5.54, and some are over 100 *m* high. The hyperbolic profile is for greater structural strength, not for any thermodynamic reason.

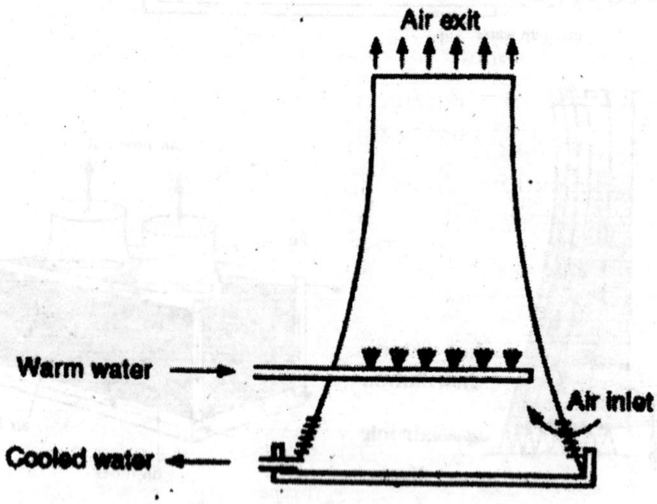

Fig. 5.54 Natural draught cooling tower

Natural draught cooling towers do not require any external power to induce the air, but they cost a lot more to built than forced draught cooling towers. |

Some of the cooling water is lost to the atmosphere in the evaporation process.

In addition to the water lost due to the evaporative cooling, there are two other causes of water loss. Drift loss results from wind carrying water away with the air. Blowdown less results from draining off and discarding a small portion of the water from the basin. This must be done at regular intervals in order to prevent continual accumulation of minerals that would otherwise occur from the evaporation and drift losses.

5.15.3 Dry Cooling Tower

Fig. 5.52 (c) illustrates a dry cooling tower. Water is cooled inside finned tubes across which atmospheric air flows. Since the tower operates as a heat exchanger, water evaporation is not involved. Dry towers require more land area than other towers and, for steam power cycle applications, typically produce higher condenser pressures that result in lower cycle thermal efficiencies. Such towers are used in areas where adequate makeup for evaporative cooling towers is not available

EXERCISE

A. Theory

5.1 Draw a neat sketch of simple steam power plant.

5.2 Draw neat sketches of non-condensing and condensing steam power plant.

5.3 Draw and explain layout of steam power plant.

5.4 Draw and explain a typical layout of a steam power plant.

5.5 Discuss pointwise site selection for steam power plants.

5.6 List the points which should be kept in mind while designing coal handling plant.

5.7 List various methods of coal handling plant and explain any one.

5.8 List various stages of coal handling plant.

5.9 Draw and explain outline of coal handling equipment.

5.10 Draw and explain layout of modern fuel handling equipment.

5.11 Draw and explain line diagram of grab bucket coal handling equipment.

5.12 Draw and explain bucket of coal handling equipment.

5.13 Draw and explain of skip hoist equipment.

5.14. Draw and explain belt conveyor fuel handling equipment.

5.15. Draw and explain screw conveyor coal handling equipment.

5.16 Draw and explain scraper conveyor coal handling equipment.

5.17 What are the principal requirements of good ash handling plant.

5.18 With the help of neat sketch, discuss the various methods of ash handling.

5.19 Draw neat sketch of ash collection and transportation.

5.20 Draw a general outline of ash handling equipment.

5.21 Draw a neat sketch of ash handling and dust collecting system.

5.22 Draw and explain bottom ash sluice conveyor system.

5.23 Draw and explain fly ash hydraulic vacuum conveyor system.

5.24 Draw and explain pneumatic conveying system.

5.25 Draw and explain belt conveyor system.

5.26. Draw and explain tubular air preheater.

5.27 Draw and explain air preheater arrangement.

5.28 Draw and explain Ljunstrom air preheater.

5.29 Draw and explain regenerative air preheater

5.30 Draw and explain pressure type filter arrangement for water treatment.

5.31 Draw and explain dimeralizing plant for water treatment.

5.32 Draw and explain degasification process for water treatment.

5.33 Draw and explain deaeration process of water treatment.

5.34 Write two most commonly used for burning of coal and explain any one.

5.35 Write advantages of stoker firing.

5.36 Draw neat sketches of overfeed stoker Underfeed stoker.

5.37 List mechanical stokers and explain any one.

5.38 Draw and explain chain grate stoker.

5.39. What is the basic difference between chain grate stoker and travelling grate stoker. Draw and explain travelling grate stoker.

5.40 Draw and explain spreader stoker.

5.41 What is the principle of retart type stoker.

5.42 Describe pulverized coal firing system.

5.43 Draw and explain primary and secondary air system used in pulverzing coal firing system.

5.44 Describe unit and central system used for preparation of coal into a fine powder.

5.45 Describing three stages used in pulverizing process.

5.46 According to speed classify pulverizers.

5.47 Describe low speed pulverizer.

5.48 Describe medium speed pulverizer.

5.49 Describe high speed pulverizer.

5.50 Draw neat sketch of ball tube mill low speed pulverizer.

5.51 Draw neat sketch of ball and race mill medium speed pulverizer.

5.52 Draw neat sketch of hammer mill or impact high speed pulverizer.

5.53 Draw and explain pulverized coal burners.

5.54 Draw a neat sketch of straight flow burner.

5.55 Draw and explain patterns of flame motion in a dry bottom furnace.

5.56 Draw and explain cyclone furnace.

5.57 Draw and explain turbulent burners.

5.58 Draw arrangement of turbulent burners on furnace walls.

5.59 Draw and explain fluidized bed system.

5.60 Draw neat sketches showing types of boiler circulation.

5.61 Explain of three design concept of water tube boilers with the help of neat sketches.

5.62 Draw and explain La Mont boiler.

5.63 Draw and explain once through steam generator concept.

5.64 Draw neat sketches of three types and explain the briefly. Of cooling towers namely

(a) Induced draft evaporative cooling tower (b) Natural draft evaporative cooling tower (c) Dry cooling tower.

5.65 Draw a neat sketch of forced draught cooling tower and explain.

5.66 Draw a neat sketch of Natural draught cooling tower and explain.

5.67 Draw a neat sketch of dry cooling tower and explain them.